CHEMISTRY AND CHEMICAL ENGINEERING IN THE PEOPLE'S REPUBLIC OF CHINA

Chou En-lai and Glenn T. Seaborg

CHEMISTRY AND CHEMICAL ENGINEERING IN THE PEOPLE'S REPUBLIC OF CHINA

A TRIP REPORT OF THE U.S. DELEGATION IN PURE AND APPLIED CHEMISTRY

Edited by John D. Baldeschwieler

1979 AMERICAN CHEMICAL SOCIETY
Washington, D.C.

Cover: A drawing exhibited in Five-Springs Park in Lanchow, visited by the delegation. It represents a petrochemical refinery at Tach'ing and is a child's attempt to depict the slogan, "Learn from Tach'ing," the oil field in Northeast China that is held up as an example of progress.

Library of Congress Cataloging in Publication Data

U.S. Delegation in Pure and Applied Chemistry.
Chemistry and chemical engineering in the People's Republic of China.

1. Chemistry—China. 2. Chemical engineering—China. I. Baldeschwieler, John D., 1933– II. Title.

QD18.C5U54 1979 540'.951 79-11217
ISBN 0-8412-0501-9 1-266 1979
ISBN 0-8412-0502-7 (pbk.)

PRINTED IN THE UNITED STATES OF AMERICA

CONTENTS

Contributors

Glenn T. Seaborg (Chairman)
Associate Director
Lawrence Berkeley Laboratory
University of California
Berkeley, California 94720

John D. Baldeschwieler (Deputy Chairman)
Professor of Chemistry
Division of Chemistry and Chemical Engineering
California Institute of Technology
Pasadena, California 91125

Jacob Bigeleisen
Vice-President for Research and Dean of Graduate Studies
State University of New York at Stony Book
Stony Brook, New York 11794

Ronald Breslow
Professor of Chemistry and Chairman
Department of Chemistry
Columbia University
New York, New York 10027

Robert G. Geyer
Staff Officer
Committee on Scholarly Communication with the People's Republic of China
National Academy of Sciences
Washington, D.C. 20418

James A. Ibers
Professor of Chemistry
Department of Chemistry
Northwestern University
Evanston, Illinois 60201

Thurston E. Larson
Principal Scientist, Emeritus
Illinois State Water Survey
Urbana, Illinois 61801

Yuan T. Lee
Professor of Chemistry
Department of Chemistry and Lawrence Berkeley Laboratory
University of California at Berkeley
Berkeley, California 94720

Alan Schriesheim
Director
Corporate Research Laboratories
Exxon Research and Engineering Co.
Linden, New Jersey 07036

Richard S. Stein
Professor of Chemistry
Department of Chemistry
University of Massachusetts
Amherst, Massachusetts 01003

Richard P. Suttmeier
Associate Professor and Chairman
Department of Government
Hamilton College
Clinton, New York 13323

James Wei
Professor of Chemical Engineering and Department Head
Department of Chemical Engineering
Massachusetts Institute of Technology
Cambridge, Massachusetts 02139

FOREWORD

We present here a rather complete description of the research in chemistry and chemical engineering and related areas observed by our 12-member Delegation in Pure and Applied Chemistry during our three-and-a-half week visit to the People's Republic of China in May–June 1978. In order to place our observations in perspective in the face of the current changing conditions in China, we also include a rather substantial account of the historical background of science and its administration in China.

Ours was, in a sense, reciprocal to a visit to the United States made almost exactly a year earlier by a group of chemists from China led by the distinguished quantum chemist T'ang Ao-ch'ing, President of Kirin University. Several members of our delegation had met and formed friendships with these visitors, and we were pleased to have the opportunity to renew our acquaintance with each of the 10 chemists from that delegation during our travels to China.

This was my second visit to the People's Republic of China. I was privileged to be a member of the delegation sponsored by the U.S. Committee on Scholarly Communication with the People's Republic of China that visited China five years earlier in May—June 1973. This earlier delegation had the responsibility for negotiating the terms under which these exchange visits are taking place, and it was under this exchange agreement that the visit of our Delegation on Pure and Applied Chemistry took place. During the visit in 1973 we had the unusual opportunity to meet with Premier Chou En-lai (Fig. 1) to discuss the conditions for future exchange visits, which would take place under the provisions of the Shanghai Communiqué of 1972. During our hour-and-a-half meeting the Premier gave us a wide-ranging view of pertinent Chinese historical background, and on this occasion told us he would accept nine out of the twelve subject areas we had requested to have covered by our exchange agreement. During this early exciting visit, I visited a number of chemical laboratories. This, of course, was during the period when the influence of the Cultural Revolution was still strongly felt and the "Gang of Four," whose existence was unknown to us, was in its ascendancy. Their position of power placed an interesting flavor on our discussions and provided contrast to the visit described in this report.

Figure 1. Chou En-lai greeting Dr. Glenn T. Seaborg at meeting in Peking in 1973.

It is impossible to describe adequately the degree of warmth and cordiality with which we were received by our Chinese colleagues throughout our visit. Their warmth did much to ease the weight of our very full schedule, made so strenuous because our hosts wanted us to see and learn as much as possible. During our visits to the laboratories, educational institutions, and industrial plants, we received the royal treatment. Our hosts in each city entertained us with a multicourse banquet, punctuated with the traditional Chinese toasts, in which mao tai was featured and which always included welcoming remarks by the evening's host at the beginning and my response near the end of the meal. The theme of these remarks always emphasized an increasing friendship between the scientists and the peoples of our two countries and often included an expression of the hope that the degree of our cooperation in science might increase.

We were impressed by the high degree of motivation exhibited by the students and young scientists. They are anxious to learn, do good research, and serve their country in whatever task they might be assigned. They were very familiar with the foreign scientific literature.

One of the interesting features of our visits to the various research institutes and universities was the opportunity for the members of our delegation to give talks in their specialties. In many instances such a talk, together with the following question-and-answer period, would consume half a day. In some cases a delegation member would give such a talk in the morning and another talk in the afternoon of the same day. We were impressed with the high quality of the questions following our talks, indicating an extraordinary knowledge of the relevant scientific literature by the members of our audience.

Of particular interest is the fact that we were the first delegation to visit the mid-China industrial city of Lanchow. This made us objects of curiosity to an even greater extent than during the rest of our visit. Following a dinner in Lanchow Restaurant, our delegation found a crowd of some 500 curious people (Fig. 2), including many children, waiting for us when we emerged from the restaurant. They were smiling, and very, very friendly, but obviously we were a strange species.

Figure 2. Crowd greeting the U.S. Delegation outside a restaurant in Lanchow.

Many interesting comparisons can be made between this visit and my visit five years earlier. The most apparent difference results from the overthrow of the "Gang of Four" and the complete termination of the Cultural Revolution. Five years ago there was much talk about the end of the Cultural Revolution in the sense that universities and colleges were apparently being reopened, and there was talk of starting graduate education again. Apparently this was a false hope, and there was a setback extending into 1974–75 due to the actions of the "Gang of Four." Every briefing at the start of our welcoming sessions made reference to conditions before and after the "overthrow." All universities are starting to take advantage of the national entrance examination, are expanding back to the four-year (from the three-year) curriculum, and all the major universities and research institutes are now starting to admit graduate students on the basis of the national examination augmented by additional local examinations. An important consequence of this reform is the new turn to basic research. We were told that emphasis on applied research was, of course, necessary during the years before the Cultural Revolution, but this would normally have been followed by a turn to basic research in the 1960s had this natural course of events not been interrupted by the Cultural Revolution and/or the "Gang of Four."

Our visit came at a propitious time from the point of view of the real changes that have occurred and are occurring following the Cultural Revolution. We suggested to our hosts in Peking, and at the other places we visited, that our two-country exchange agreement be expanded to include bilateral symposia on selected subjects, exchange of graduate students and other scientists for extended periods of time, and other means of increasing cooperation. We had the impression that our hosts have a great desire to move in this direction, and events that have occurred since our return home, including the normalization of relations between our countries, give me much encouragement that we shall indeed see much more of our Chinese friends.

Any accurate and complete assessment of the impact of our visit must, of course, await future developments. Our goals include making known to our colleagues in the United States some of the vast amount of chemical information that now resides in the People's Republic of China. We hope that this report will serve a useful role toward this objective and that our colleagues will have their appetites whetted to the point that they will seek further information. We have deliberately included a large number of names to serve as points of contact for this purpose. Another goal is the expanded transfer of chemical information from our colleagues in the United States to our vastly increasing circle of friends in the People's Republic of China. These two aims are, of course, closely interrelated as components of the traditional mechanism for the exchange of scientific information. We believe that such increasing, friendly cooperation in scientific fields, in view of the advantages to both our countries, has been and will continue to be a positive force toward an improving political relationship.

As a final word I might add that all of our 12-member delegation, many of whom met each other for the first time on this journey, managed to remain on good terms throughout our full and arduous schedule. Even beyond that, I believe that a bond remains among the members of our group as a pleasant dividend ensuing from our interesting, mutual adventure. And we all acknowledge a debt of gratitude to John Baldeschwieler, our Deputy Chairman, for the competent manner in which he has served as Editor to meld the various contributions of our delegation members into this, hopefully, coherent report.

December 20, 1978

Glenn T. Seaborg
Berkeley, California

PREFACE

The U.S. Delegation in Pure and Applied Chemistry visited China in May and June 1978 under the auspices of the Committee on Scholarly Communication with the People's Republic of China (CSCPRC), a group jointly sponsored by the National Academy of Sciences, the Social Sciences Research Council, and the American Council for Learned Societies. The topic of pure and applied chemistry was first suggested by the CSCPRC in June 1976 during negotiations for the 1977 exchange program with China, whereupon the Chinese Science and Technology Association (the Committee's counterpart organization) proposed a visit of a group of Chinese chemists to the United States. As a result of this agreement, the Committee hosted a delegation of Chinese chemists in April and May 1977 (Fig. 3). During their

Figure 3. Members of the Chemistry Delegation from China visiting in Berkeley, California, May 26–27, 1977 from left to right: Su Feng-lin, Ms. Chou En-lo, Wang Erh-k'ang, Ms. Hsia Tsung-hsiang, Wu Yueh, T'ang Ao-ch'ing, Huang Wei-yuan, Chang Lo-feng, Chiang Ping-nan, Yin Yuan-ken, Hsu Mao.

visit, which took them to many universities, companies, and government laboratories, the Chinese met many American scientists, some of whom participated in the return visit of the U.S. Delegation in Pure and Applied Chemistry.

With little previous knowledge of chemistry in China, we planned the trip relying on published accounts of research in chemistry and chemical engineering, on consultation with several chemists who recently had visited China, and on suggestions provided by the Chinese delegation that had visited the United States. Taking as a base the institutions represented by the Chinese group, we drew up a suggested itinerary that was sent to the Chinese Scientific and Technical Association in April 1978. All of these requests were met by our Chinese hosts.

The trip began in Peking with visits to the Institute of Chemistry of the Academy of Sciences, as well as to other leading research institutes in chemistry, physics, environmental chemistry, pharmaceutical chemistry, and biophysics. The delegation then divided into two subgroups, one visiting Harbin and the Tach'ing oil field complex in Northern Manchuria, while the second subgroup visited the Institute of Chemical Physics and nearby industrial installations in Talien. The delegation reassembled in Ch'angch'un to visit the Kirin Institute of Applied Chemistry and Kirin University, and then it proceeded to Shenyang and Fushun to visit a number of the industrial activities in that area. The delegation visited a number of research institutes and industrial activities in chemistry and chemical engineering in the Shanghai area, including an excursion to the attractive city of Hangchow. A photograph of the U.S. Delegation in Hangchow is shown in Fig. 4. From Shanghai we traveled to Sian to visit Northwest University, as well as the recently discovered archaeological site on the outskirts of Sian. From Sian, the Delegation visited research institutions in the city of Lanchow on the Yellow River in the central part of China. A map of China showing the itinerary of the U.S. Delegation is shown in Fig. 5, while the details of the itinerary appear in Appendix B, and a list of our Chinese hosts can be found in Appendix C to this report.

We were accompanied during our visit by five very able Chinese, Dr. Ch'ien Jen-yuan, Deputy Director of the Institute of Chemistry in Peking (Fig. 6), Dr. Shih Liang-ho, a polymer chemist from the Institute of Chemistry in Peking, Mr. Teng Shao-lin, Librarian of the Peking Institute of Chemistry, Mr. Shih Wei-ming, Peking Institute of Chemistry, and Mrs. Hu Feng-hsien, Ministry of Foreign Affairs.

High quality research and development work in chemistry and chemical engineering is often carried out in "nonchemical" institutions and is not necessarily to be found only in institutes of chemistry

Figure 4. *Members of the U.S. Delegation in Pure and Applied Chemistry at the Hsi-ling Guest House, Hangchow, June 4, 1978. From left to right: John Baldeschwieler, James Ibers, Jacob Bigeleisen, Alan Schriesheim, James Wei, Richard Suttmeier, Ronald Breslow, Glenn Seaborg, Yuan Lee, Robert Geyer, Thurston Larson, Richard Stein.*

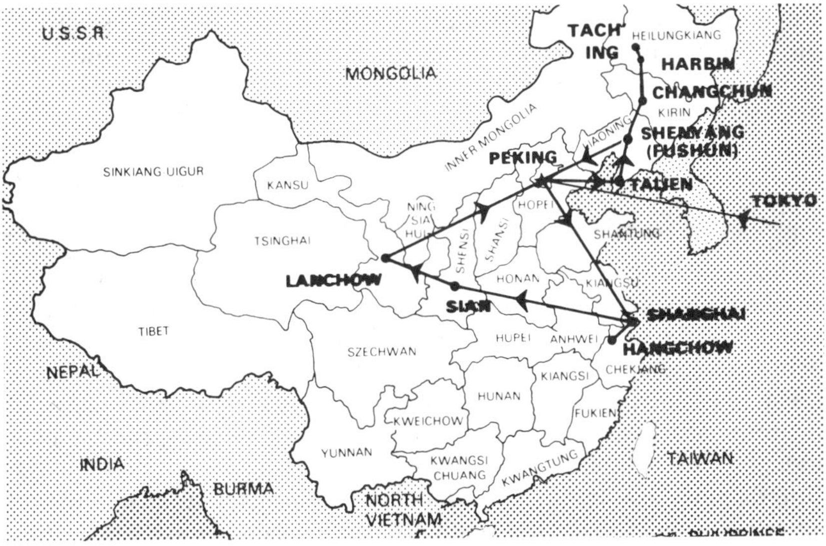

Figure 5. *Map showing cities visited in China by the U.S. Delegation in Pure and Applied Chemistry.*

Figure 6. Ch'ien Jen-yuan.

or chemical engineering, or under the Chinese Academy of Sciences. The Ministry of Petroleum Industry, the Ministry of Chemical Industry, the Ministry of Public Health of the Chinese Academy of Medical Science, and the Ministry of Education are all involved. For example, some good physical organic chemistry and some of the best instrumentation was reported to be found in the Institute of Photography in Peking. Excellent work in solid-state chemistry, physics, and materials science is found in the Institute of Ceramic Chemistry and Technology, and we were told that research in isotope separation and inorganic chemistry is carried out in the Institute of Saline Lakes in Hsining. Although the name of the Institute of Applied Chemistry in Kirin Province appears to stress applied work, some of the best basic research in chemistry that we observed is carried out there. At the Institutes of Chemical Physics in Talien and Lanchow, the focus of research was not on chemical physics as normally defined in the United States (spectroscopy, molecular structure, quantum chemistry) but rather on more classical methods of physical chemistry including thermodynamics, kinetics, and polymer science.

Institutes that the delegation probably should have visited had time and prior information been available would certainly include the Institute of Photography (Peking), the Institute of Saline Lakes (Hsining), the Chemical Engineering Institute of the University of Tientsin, the University of Science and Technology (Hofei, Anhui Province), the Institute of Rare Earths (Anhui Province), the Institute of Metallurgy (Shanghai), the Institute of Chemical Engineering and Metallurgy Research (Peking), and Szechwan University (Chengtu).

The mission of the U.S. Delegation in Pure and Applied Chemistry was to become acquainted firsthand with the status of research, development, and teaching programs in chemistry (not including biochemistry) and chemical engineering in the People's Republic of China. In addition, we hoped to establish personal contact with working scientists and science administrators in China to provide a base for future communication between the Chinese and the U.S. chemistry and chemical engineering communities.

We found that we were able to discuss the possibility of cooperative programs with our Chinese colleagues in a number of areas through mechanisms such as joint symposia, visiting scholars, and joint research programs. Our Chinese hosts were very receptive to these ideas, and the recommendations of this report include a number of specific suggestions for advancing such continuing communication. With the visit of the first official U.S. Science and Technology Delegation to China led by Dr. Frank Press, the President's Science Advisor, which occurred shortly after the return of our delegation; a return visit in October led by Dr. Chou Pei-yuan, President of the Science and Technology Association and Vice-President of the Chinese Academy of Sciences; and the establishment of formal diplomatic relations with China in December 1978, the climate for implementing the recommendations contained in this report is very good. The delegation was indeed fortunate to have had the chance to visit China at such a propitious time, and we believe that our discussions had a significant impact on the role of science and technology in the evolving relationship between China and the United States.

The members of the U.S. Delegation in Pure and Applied Chemistry are prepared to assist in implementing the programs identified in this report, which we feel will be of mutual interest to U.S. and Chinese chemists and chemical engineers.

December 26, 1978

John D. Baldeschwieler
Pasadena, California

1. ROOTS OF CHEMICAL RESEARCH AND DEVELOPMENT IN CHINA

Although the history of chemistry and science more generally in China is long and impressive, the history of modern science is closely linked to China's interactions with the West beginning in the nineteenth century. It was not until the twentieth century, however, that organized modern research began in earnest. An initial impetus to the development of modern science was the formation in 1914 of the Science Society of China, organized by students who had studied in the United States.

During the Republican period new research organizations and professional societies were established. Notable among these were the nine institutes comprising the Academia Sinica in Nanking and six institutes of the Peking Academy. Within the Academia Sinica there was an Institute of Chemistry established in 1928 in Shanghai. The Institute moved to Kunming during World War II and later returned to Shanghai. The Peking Academy also had an Institute of Chemistry, which was established in 1929 and which also went to Kunming during the war. In all, some 218 research installations were established in China before 1950.

The Chinese Chemical Society, which published a journal of chemistry, was founded in 1934 and reportedly had approximately 2000 members by the late 1930s. In all, some 57 scientific societies had been established before 1950.*

The Institutes of the Academia Sinica and the Peking Academy served as nuclei for a new Chinese Academy of Sciences (CAS) established after the Communist victory in 1949. The new government was strongly committed to science and technology, and it set about establishing a significantly expanded network of scientific and technological institutions. Throughout the 1950s, the influence of the Soviet Union on this institution's building strategy was strong. It resulted in a division of Chinese science and technology into three main sectors: (1) an academy sector headed by a strong centralized Soviet-styled Academy of Sciences; (2) a ministerial sector in which production ministries established their own research and design

* For pre-1949 background, see Yuan-li Wu and Robert Shecks, "The Organization and Support of Scientific Research and Development in Mainland China," Praeger: New York, 1970.

institutes; and (3) a higher education sector. Work in chemistry therefore was done not only in selected institutes of the CAS but also in various ministerial institutes and in universities.

In 1956 the Chinese began research planning in earnest. A 12-year plan was formulated, specifying not only research projects but also new fields of study to be initiated, facilities to be built, and manpower to be trained and deployed. The plan identified the following areas of work as having the highest priority: (1) peaceful uses of atomic energy; (2) radio electronics; (3) jet propulsion; (4) automation and remote control; (5) petroleum and scarce mineral exploration; (6) metallurgy; (7) fuel technology; (8) power equipment and heavy machinery; (9) the harnessing of the Yellow and Yangtze Rivers; (10) chemical fertilizers and agricultural mechanization; (11) prevention and eradication of detrimental diseases; (12) problems of basic theory in natural science.

In 1958 a number of important changes in Chinese science and technology occurred. First, a powerful, centralized Science and Technology Commission (STC) was established as the chief science policy and administrative agency of the country. The STC was to coordinate the three sectors and was to lead in the plan development and implementation processes. Second, the All-China Federation of Scientific Societies and the All-China Association for the Promotion of Scientific and Technical Knowledge were merged into the Science and Technology Association (STA). As the elements of the merger suggest, the STA was to promote the linking of professional science to the popularization of science. Finally, in keeping with the themes of the "Great Leap Forward," political attention focused on science and technology at the grass roots of society and on ways of mobilizing mass enthusiasm and participation in science-related activities.

This latter development, which marks the beginning of what is sometimes called "walking on two legs" in science (professional and mass), contained a number of different elements. First, efforts were made to establish research activities under the control of the provincial governments. Accordingly, most provinces established "branch academies" of the central Academy of Sciences. The provinces also established scientific and technological committees and scientific and technological associations to parallel the central STC and STA. Second, much media attention was given to peasant and worker scientists and innovators, and the "wisdom of the masses" was widely extolled. Finally, aspects of professionalism in science were criticized, particularly those aspects suggesting elitism in science and scientists' preoccupation with research unrelated to practical problems.

In attempting to cope with the excesses of the Great Leap Forward, these latter themes were somewhat muted in the 1961–66 period. Similarly, some of the more unworkable efforts at establishing decentralized scientific institutions were abandoned.

By 1965 Chinese science and technology had experienced impressive growth and development. The Academy of Sciences, for instance, had grown from around 20 institutes in 1950 to well over 120 in 1965. Whereas Chinese institutions of higher learning were graduating some 9300 scientists and engineers (including agriculture, forestry, and medicine) per year in 1949–50, the number of annual graduates had swelled to 129,000 by 1962–63.* And in terms of expenditures the science budget increased enormously, with expenditures increasing from an estimated one million yuan in 1950, to an estimated 1.47 billion yuan in 1963.† This pattern of quantitative growth, however, was severely disrupted by the Cultural Revolution that began in 1966.

The full story of the impact of the Cultural Revolution on science and technology is yet to be told. A few generalizations, however, can be ventured. First, the antiprofessionalism themes that first appeared in the 1958–60 period returned with a vengeance. Suggestions of elitism were severely criticized, and narrowly conceived standards of relevance and practicality in research were enforced. An additional manifestation of antiprofessionalism was the suspension of the activities of such professional societies as the Chinese Chemical Society and the Chinese Society for Chemical Engineering (founded in 1957).

Second, major changes were made in the institutional structure for research. The STC was abolished, approximately three-fourths of the institutes under the CAS were removed from CAS jurisdiction and placed under the jurisdictions of local governments, and the system of having professional leadership in research institutes was replaced with leadership by revolutionary committees comprising professionals, representatives of the masses, and Party cadres.

The most serious effect of the Cultural Revolution on science was probably the impact it had on higher education. Universities were shut down during the height of the Cultural Revolution, and when they were reopened, they were a shadow of their former selves. Thus, China now enters her drive for the comprehensive modernization of the country by the year 2000 not only with a number of research opportunities missed as a result of the Cultural Revolution but, more seriously, with a missing age cohort of scientists resulting from 10 years of disrupted higher education and advanced training.

* Leo A. Orleans, "Scientific and Technical Manpower," in *Science and Technology in the People's Republic of China*, OECD: Paris, 1977.

† Wu and Shecks, p. 194.

The final generalization about the Cultural Revolution is that generalizing must be done with care. That is, it is not clear that the impact was as great in some areas and fields as in others. While we heard frightening stories about the treatment of scientists during the Cultural Revolution, we also saw evidence of progress at some places that could only have been made in the absence of major disruptions.

The formal end of the Cultural Revolution is normally put at 1969. Many institutional "reforms" from the Cultural Revolution persisted beyond that date, however, and continued to affect science and technology (e.g., the insistence that scientists should go to the countryside and engage in physical labor). In addition, the period from 1969 to the end of 1976 was one of high level political and organizational instability throughout the society, as those who had been removed from office during the Cultural Revolution began to reappear and vie for positions with those whose careers benefited from the Cultural Revolution. These disagreements about the desirability of continuing reforms begun during the Cultural Revolution were complicated by career interests at various levels in the society. At the highest political levels, the instability was directly linked with the question of political succession.

Science and technology became deeply embroiled in this struggle after 1972. At that time Mao Tse-tung and Chou En-lai reportedly became convinced that science and technology and advanced education had to be strengthened and that more attention had to be given to basic research. Chou solicited the views of the leaders of the scientific community, but before their recommendations could be published and advanced, they were allegedly sabotaged by the group of political leaders that has come to be known as the "Gang of Four."

The tension between the Gang of Four and those urging a major new effort in science and technology broke out in 1975. In January of that year, Chou En-lai called for the comprehensive modernization of agriculture, industry, national defense, and science and technology by the year 2000. In one of the initial moves to implement this policy, a major review of the state of science was undertaken, a review that became known as the "Outline Report." The appearance of the Outline Report elicited a strong response from the Gang of Four, who alleged that, if implemented, the Outline Report would return China's scientific policy and organization to the standing of 1965, and would "reverse the verdicts" of the Cultural Revolution.

From late 1975 to Mao's death in 1976, the views of the Gang of Four prevailed in the public media they controlled. With the removal of the Gang of Four from office in late 1976 and the reinstatement of Teng Hsiao-p'ing, who had apparently initiated the study leading up to the preparation of the Outline Report, changes in scientific policy have been rapid and dramatic. Throughout 1977 a series of conferences both of Party officials and of scientists were held. The former resulted in the making of a national political commitment to rapid scientific and technological development. The latter encouraged the reawakening of professionalism in science and marked the beginning of preparations for a new national plan for scientific development. The preparations of 1977 culminated in the National Science Conference of March 1978 at which scientific achievement was recognized and the broad outlines of the plans for future development were announced.

2. THE INSTITUTIONAL STRUCTURE OF CHEMICAL RESEARCH AND DEVELOPMENT IN CHINA

Since October 1976 when the Gang of Four was ousted, scientific and technological policy and organization have been subjects of intense national review and discussion. Although many parts of the institutional structure for research and development are now in place, a number of institutional changes are still in process. Nevertheless, the broad outlines of organization are now evident, and these show important similarities to the system of organization that existed before the beginning of the Cultural Revolution in 1966.

The national institutional structure for research and development, hereafter China's "science system," is characterized by vertical and horizontal divisions. Vertically, the science system is divided into three main centrally controlled sectors: The Chinese Academy of Sciences (CAS) sector; the university sector led by the Ministry of Education; and the production ministry, or ministerial, sector. Control of and coordiantion among these sectors on matters relating to science and technology is the responsibility of the recently reestablished Science and Technology Commission.

Efforts at administrative decentralization that go back to the 1958–60 period have resulted in a horizontal division of the science system as well. Administrative authority is shared principally by the central government and by provincial governments.* However, there is also a horizontal layering of administrative authority at the subprovincial level of cities, counties, communes, production brigades, and production teams. This horizontal division has resulted in many research and production units being under the dual control of the center and the province, while others are principally under provincial control.

In short, the science system is variegated and complex. Our understanding of its operation is incomplete, but some of the examples cited in what follows will illustrate aspects of its operation.

* The term "provincial government" includes as well the three major urban centers of Peking, Shanghai, and Tientsin.

The Science and Technology Commission

The center for research policy, planning, and national coordination is the Science and Technology Commission (STC), which was reestablished in 1978 after its dismantling during the Cultural Revolution. The delegation did not have an opportunity to meet with representatives from the central STC, and we were told that it is still in the process of being organized. The STC is headed by Fang Yi, a nonscientist, and formally Vice-Minister for Economic Relations with Foreign Countries. Fang, at present, also holds positions as a Vice-Premier and a Vice-President of CAS. Most important, however, he is a member of the Communist Party's Political Bureau, the pinnacle of political power in China. The fact that a member of the Political Bureau heads the STC is interpreted as an indication of the depth of political commitment made to scientific development following the ouster of the Gang of Four.

The detailed organization and operation of the STC is not known. It is known to have a Planning Bureau (headed by Yung Lung-kui), a bureau dealing with national energy research (No. 2 Bureau headed by Lin Hua), and a bureau dealing with industry, communications, and new technology (No. 3 bureau headed by Yeh Hsuan-p'ing). It is thought that the STC also has a basic research bureau. The Vice-Ministers of the STC include experienced science administrators Yü Kuang-yuan and Wu Heng, Chiang Nan-hsiang, a former president of Tsinghua University and former Minister of Higher Education (1965–66), Chao Tung-wan, T'ung Ta-lin, and Chiang Ming.

The Chinese Academy of Sciences

The central administrative offices of CAS, usually referred to as "the Academy of Sciences," are in Peking. Under the Academy are research institutes that are distributed around the country in all the natural sciences.* As a result of the Cultural Revolution, the number of institutes directly under the CAS was drastically reduced from more than 120 to 36. CAS jurisdiction is again being established over many institutes, and although the process is not yet complete, we were told that CAS jurisdiction now extends to some 90 institutes including those where jurisdiction is shared with other administrative units.

The pre-Cultural Revolution system of having regional "branch academies" is being gradually reestablished. The Shanghai Branch is now in operation and is headed by biochemist Wang Ying-lai, who is also the Director of the Institute of Biochemistry. Reportedly provin-

* Before the Cultural Revolution, CAS also contained social science institutes. These have recently been gathered under a newly created Academy of Social Sciences.

cial branch academies have been set up in Szechuan and Sinkiang, and other provinces are planning their establishment.

The longtime President of CAS, Kuo Mo-jo, who held the position since 1949, died shortly after the Delegation left China and a new president has not yet been named. The Vice-Presidents are, in addition to Fang Yi, scientists Chou P'ei-yuan, T'ung Ti-chou, Yen Ch'i-tze, Ch'ien San-ch'iang, Hua Lo-keng, and nonscientists Li Ch'ang and Hu Ko-shih.

Policy making, planning, and coordination within the CAS sector is centered in the Academy's Secretariat led by the powerful office of the Academy's Secretary-General. The present Secretary-General is an experienced science administrator, Yü Wen. Within the Secretariat are five discipline-oriented bureaus, one each for biology; mathematics, physics, and astronomy; "new technology"; chemistry; and earth sciences. These bureaus are the points of contact for institute leaders and are the centers for major allocative decisions affecting research within the Academy sector. Bureau staffs reportedly consist of scientists as well as nonscientists. The Academy is in the process of organizing "academic committees" as advisory mechanisms at the bureau level. These will be composed of leading scientists within a given discipline from both within and outside the Academy sector. For major projects involving substantial expenditures, the bureaus also convene *ad hoc* advisory committees composed of scientists, and where appropriate, engineers and representatives from industry.

Chemistry within the Chinese Academy of Sciences is conducted in 14 research institutes under the Academy's Bureau of Chemistry. These are the following:

In Peking:	Institute of Chemistry*
	Institute of Environmental Chemistry*†
	Institute of Photography†
	Institute of Chemical Engineering and Metallurgy
In Shanghai:	Institute of Organic Chemistry*
	Institute of Ceramic Chemistry and Technology*
In Talien	Institute of Chemical Physics*
In Lanchow:	Institute of Chemical Physics*
In Ch'angch'un:	Institute of Applied Chemistry*
In Foochow:	Institute of Structural Chemistry
In Tsinghai Province:	Institute of Saline Lakes
In Ch'engtu:	Szechuan Institute of Chemistry
In T'aiyuan	Institute of Coal Chemistry
In Anhui Province:	Institute of Rare Earths‡

* Indicates institutes visited by delegation members.

† Formerly part of the Institute of Chemistry; newly established or separate institutes.

‡ Formerly part of the Institute of Applied Chemistry; recently established as separate institute.

In addition to institutes doing work in cognate fields to chemistry (physics and biochemistry, which are under other bureaus), American chemists will also be interested in the existence of a CAS Institute for Scientific Information and an Academy-directed scientific instrument design facility, both of which are in Peking and are believed to be under the jurisdiction of the Secretariat.

CAS institutes vary in size from 300–400 employees to over 1000 employees. Because the institutes offer housing, health care, child care, and other services to their employees, it is not unusual to have as many as one-third of the personnel engaged in administrative tasks not directly related to research.

After the Cultural Revolution, leadership in CAS institutes was in the hands of "revolutionary committees." Following the purge of the Gang of Four, revolutionary committees have been dissolved, and leadership is again being organized in terms of a director (usually a scientist) and deputy directors (some of whom are usually scientists). For research purposes, institutes are subdivided into laboratories (typically about five in number), which in turn are divided into research groups and teams. For research planning and administration, the institute will have one office for equipment and one for research administration. These offices are usually led and staffed by people with some scientific training. Of particular interest are the activities of the research planning and administration office. It is responsible for developing research plans for the institute, for administering the plans, for coordinating the plans with other institutes and organizations, and for training research personnel. Working with the laboratories, it develops planning recommendations. These recommendations are then passed on to the newly reestablished institute-level "academic committee" for deliberation and decision. Institute academic committees are composed of 20–30 leading scientists and high-level administrators.

Policy-making and planning within the Academy sector then is characterized by downward and upward communication. Research projects often come to the working scientist from above as part of a larger research plan that in turn is linked to economic plans. During the 1966–76 period, heavy emphasis was put on applied research. Since the purge of the Gang of Four, basic research has gained legitimacy, and the CAS, along with the universities, will become the chief performers of basic research in the future. It seems likely therefore, that in the planning process, considerably more initiative for plan development will come from the laboratory and institute levels.

Although the delegation generally did not have access to information concerning budgets, we did learn that institutes usually operate under three types of annual budgets: a general purpose budget, a foreign exchange budget, and a manpower allocation. There are evident limitations of foreign exchange and a shortage of trained manpower.

Does such a planned and highly organized system have the capacity to respond favorably to creative new ideas of an individual scientist? Members of the delegation raised this question on a number of occasions and uniformly received affirmative answers. Worthy proposals that require moderate amounts of resources can be accommodated at the institute level out of normal budgetary allowances. More ambitious proposals that go beyond current budgets are taken to the central Academy for review, and if they have merit, are incorporated into future plans. Since budgets seem to be expanding, the Chinese planning/budgeting systems seem to offer the prospects for predictable and reliable research support.

The University Sector

Of all the sectors of Chinese science, perhaps the one most disrupted by political events during the past 12 years was the university sector. As in other sectors, the research that was done during this period was largely applied, intended to serve rather immediate production needs. As a result of policy changes during the past two years, universities, like the CAS, are to devote more effort in the future to basic research.

Institutions of higher education in China can be divided into the following categories:

1. Comprehensive universities, or those offering courses in a broad spectrum of disciplines in the arts and sciences. Peking University, Kirin University, Futan University, and Northwest University, visited by members of the Delegation, fall into this category.
2. Universities specializing in the sciences or engineering. Tsinghua University, seen by members of the delegation, falls into this category.
3. Specialized institutes for particular areas of technology. The Shanghai Institute for Chemical Engineering, visited by Professor Wei, is an example of this type of institution.
4. Worker universities and schools run by production units. This latter category was an innovation of the Cultural Revolution and reportedly will be kept.

In addition to this categorization, higher education can also be differentiated on the basis of a newly established program for developing centers of excellence. Eighty-eight institutions have been so identified. In addition, three universities—the University of Science and Technology in Hofei, Anhui Province; Chekiang University, Chekiang Province; and a new University of Science and Technology in Harbin—are operated jointly by the Ministry of Education and the CAS.

Most institutions of higher education are under the direction of the Ministry of Education. Many, however, also have strong ties to local government as well, particularly with regard to research. Other sections of this report discuss university research and education in some detail. The remainder of this section will outline the institutional structure for university research based on the site visits made during the trip. One must be cautious, however, in generalizing these cases because the particulars of different localities create variation.

Most universities receive an annual research budget from the Ministry of Education. This budget includes funds for research and for facilities. Like the CAS sector, research is conducted according to a plan that is centrally coordinated. The Ministry reportedly has an office of science and technology that is responsible at the national level for university research. There is also an education bureau at the province level that has an office for science and technology. In addition, a university may receive projects and funding for applied research from the central Science and Technology Commission and its provincial branches. Finally, research projects that are initiated by university scientists but are outside of the year's plans can be funded at the university level out of discretionary funds available there, or application can be made for support through either the provincial education bureau or the central Ministry of Education.

In the past, the universities, unlike CAS, performed contract research. This reportedly was due to the fact that university research budgets were not as bountiful as those of CAS. It is possible that university research budgets were kept low in order to encourage contract work with industry and to keep research in the educational sector oriented towards "serving the needs of production." With the new emphasis on basic research and the designation of the universities and CAS as the chief performers of basic research, it is likely that central funding through the Ministry of Education and scientist-initiated research activities will increase.

A variation on this pattern is the situation at Northwest University in Sian. The university's main budget comes from the Ministry of Education via the province's education bureau. However, some of the research funds for the university come from the provincial science and technology committee, even though the relationship between the university and the science and technology committee is usually mediated by the provincial education bureau.

The Ministerial Sector

A considerable amount of research, particularly applied research, is done in institutes under the jurisdiction of production and other ministries. Our delegation visited institutes connected with the Ministries of Public Health, the Chemical Industry, and the Petroleum Industry.

Within the Ministry of Public Health is the Chinese Academy of Medical Sciences (CAMS). CAMS has approximately 13 institutes under it. They are located mainly in Peking and Shanghai. These institutes do both basic and applied research and maintain linkages both with major hospitals and with the pharmaceutical industry.

In the industrial ministries we again see a complicated picture of central and provincial administrative arrangements, both for plants and factories and for research institutes. Examples from the Ministries of the Petroleum Industry and the Chemical Industry will illustrate this complexity.

Members of the delegation visited petrochemical complexes in Peking, Tach'ing, Talien, Fushun, Shanghai, and Lanchow. The complexes at Peking, Tach'ing, and Shanghai are in some ways like small cities because, in addition to production facilities, they contain housing, hospitals, schools, and shops. These complexes are under both central and local control. Production planning and technical development are usually the responsibility of the central authorities. Included in this latter category are plant design performed by central ministry design institutes and technology acquisition (including foreign technology). Many of the administrative tasks of running the nonproduction services of the complex, however, seem to be in the hands of the local authorities. As a later section of this report will indicate, local authorities also have an important role in plant personnel management.

Within the petrochemical industry are research institutes that are controlled both by the central government and by local government. The former are linked through the ministries to major centrally controlled plants. The administration of oil fields and large refineries and petrochemical plants is in the hands of the center (except for the administration of services, noted above). Thus, a major institute such as the Peking Petrochemical Research Institute described later in this report formally is under the direction of the Director of the Peking Petrochemical Works. However, it receives its research plans from the Ministry of the Petroleum Industry and goes to that ministry for approval of scientist-initiated projects. Such an institute will also cooperate with centrally controlled CAS institutes. We were, for instance, given the example of basic research on catalysts done by the

CAS institute for Chemical Physics in Talien being transferred to the Peking Petrochemical Research Institute for applied research and development.

A slightly different pattern of centrally controlled research is seen in the research work carried on at the Tach'ing oil field located in Northeast China far from a major urban center. Research work at Tach'ing is under the overall control of the Tach'ing Science and Technology Committee, which in turn is under the Office of Science and Technology of the Ministry of Petroleum, as well as the national Science and Technology Commission and the Heilungkiang Provincial Science and Technology Committee. Under the Tach'ing STC are:

1. The Tach'ing Academy of Science and Design with more than 3000 staff (the original staff coming from the Peking Petrochemical Research Institute)
2. Mission-oriented research institutes unaffiliated with the Tach'ing Academy
3. Trouble-shooting research teams (see Fig. 7)

As Fig. 7 indicates, the main linkage between Tach'ing's research activity and its environment is the linkage with the Office of Science and Technology within the Ministry of Petroleum. Of course, the Tach'ing STC is also under the overall direction of the Tach'ing Revolutionary Committee, which is responsible for the local administration of the field and its petrochemical complex and nonproduction services.

The organization chart also illustrates the second type of industrial research organization, that controlled largely by the provincial government. The Harbin Petrochemical Research Institute is an example. This institute was originally established as an institute of CAS. As part of the decentralization that occurred after the Cultural Revolution, it was transferred to the Heilungkiang provincial authorities instead. Although it is located in the same province as the Tach'ing oil field and although it has done research relating to Tach'ing's problems, the main mission of the Harbin Petrochemical Research Institute seems to be research relevant to the provincially controlled petrochemical industry.

The discussion above illustrates the administrative complexity of China's science system. The sectoral distribution of research institutes and the two levels of administrative control result in a highly variegated pattern of R&D institutions. The system requires coordination, and the reestablishment of the STC and its local branches therefore represents one of the more important institutional reforms in this post-Gang of Four era.

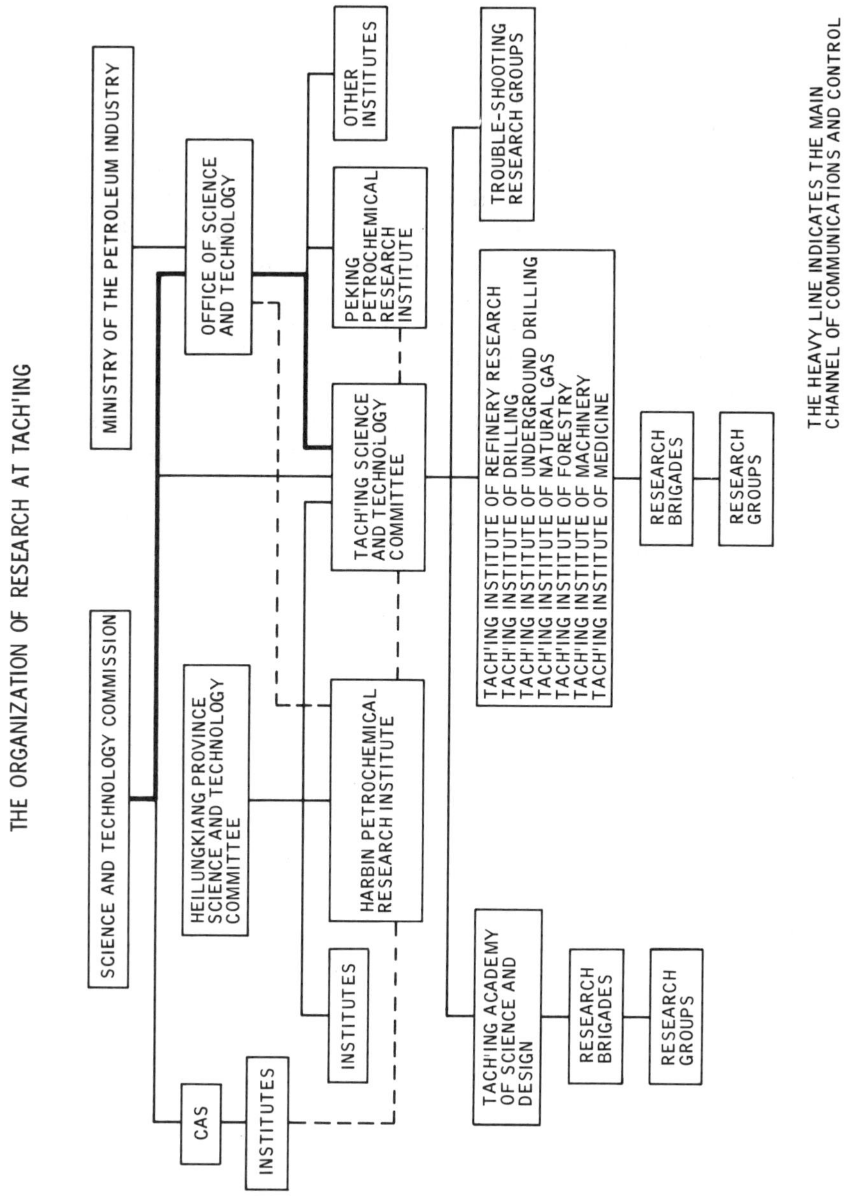

Figure 7. Organization of research at Tach'ing.

To complete this section on the institutional structure of Chinese R&D, some mention should be made of professional organizations. These too have an important role to play in integrating the science system since they open up horizontal, intersectional communications.

Before the Cultrual Revolution there were approximately 53 professional societies grouped under the leadership of the Science and Technology Association (STA). Although the STA continued during and after the Cultural Revolution, professional societies generally did not. Scientific meetings under society sponsorship were not held for over 10 years, and academic journals published by the societies ceased publication.

At present, scientific societies are being reestablished, including the Chinese Chemical Society, which convened a national meeting in September 1978. The work of the STA is to be strengthened both in serving as the peak organization for the nation's professional societies and as the leading organization for the popularization of science. The STA continues to be led by physicist Chou P'ei-yuan as President. P'ei Li-sheng, a leading pre-Cultural Revolution science administrator with the CAS, has been named the STA's Secretary-General.

3. CHEMISTRY AND CHEMICAL ENGINEERING IN THE CONTEXT OF NATIONAL SCIENCE POLICY

Science policy in China has undergone enormous changes during the past year. These changes involve new efforts to specify national goals and to relate science plans to them as well as efforts to overcome the consequences of the intense political intervention in science during the previous 10 years.

One can date the beginnings of these new directions from the Fourth National People's Congress held in January 1975. At that time, Premier Chou En-lai advanced the doctrine of the "four modernizations," calling for the comprehensive modernization of agriculture, industry, national defense, and science and technology by the year 2000. During the first half of 1975, studies were made and position papers and plans were drawn up to prepare for the implementation of the "four modernizations" policy. These plans—for both economic development and science—began to encounter strong resistance from the Gang of Four during the final months of 1975, resistance that intensified after Chou En-lai's death in January 1976. After the Gang of Four was removed from public life in October 1976, the leadership turned again to the modernization proposals of 1975 as the inspiration for current policy.

The full significance of the changes in policy became evident during 1977. Science and technology increasingly were spoken of as the keys to the realization of the other three modernizations. Scientists and technologists, who had been criticized and denounced by the Gang of Four, began to receive highly respectful treatment in the mass media. More important, the central political authorities understood and were prepared to adhere to the new line in science and technology, a line that would reverse many of the policy themes of the previous 10 years, and one that would significantly alter the environment for China's R&D.* These policy changes from 1977 were summarized as a 12-point program to strengthen science in a speech

* For a discussion of efforts to get a national political consensus on the new policy for science and technology, see Thomas Fingar and Genevieve Dean, *Developments in PRC Science and Technology Policy, October–December 1977*. S&T Summary #5, United States–China Relations Program, Stanford University.

given by Fang Yi at the Chinese People's Political Consultative Conference in December 1977.†

The speech included a critical assessment of the weaknesses in China's science system and a frank acknowledgment of the damage done by the Gang of Four. It then went on to announce the following policy actions:

1. The Science and Technology Commission was reestablished in September 1977 as the central planning and coordinating body for all scientific and technological work. The Commission had been dismantled during the Cultural Revolution. Early in 1978 Fang Yi was named the Director of the Commission with ministerial status.
2. Reforms are being made in the leadership of research institutes and universities. The reforms include the purging of followers of the Gang of Four, and the abolition of the "revolutionary committee" form of leadership organization (a Cultural Revolution innovation) and reinstatement of professional scientists as research institute directors.
3. Planning for science and technology has resumed. The plans are intended to map out the comprehensive development of science and technology by the year 2000 and involve a 3-year plan to 1980, an 8-year plan to 1985, and a 23-year plan to 2000.
4. Readjustments are being made in administrative jurisdictions, and research institutes that had been disbanded are being reestablished. During and after the Cultural Revolution a number of institutes were either disbanded or, in effect, converted to factories. In addition, many institutes were placed under provincial jurisdiction. In the interests of planning, coordination, and control, a greater degree of administration centralization is being reintroduced.
5. Efforts are being made to rationalize the utilization and employment of scientists and engineers. The Gang of Four and the Cultural Revolution more generally had an adverse effect on the morale of professional personnel. Scientists and engineers were politically suspect by the Gang of Four and were often seriously underemployed in positions not requiring the knowledge and skills they possessed. This fifth point of the program is intended to

† Fang Yi, "Report on China's Science and Education," New China News Agency, December 29, 1977. In Foreign Broadcast Information Service, December 30, 1977.

rectify this situation by reassigning personnel to more appropriate positions and also to reinstate academic titles and to recognize professional achievements.

6. Academic conferences and professional meetings are being encouraged. Again, these were seriously disrupted during the previous 10 years. In particular, the work of professional societies was halted. This is all being changed. Professional societies are again becoming active, and even before Fang's speech, a number of meetings were being held.
7. The system of university and college enrollment is being reformed. This topic is discussed elsewhere in this report, but two points should be emphasized here. First, a much greater emphasis is being put on achievement and academic excellence instead of political criteria in recruiting individuals for higher education. Accordingly, it is not mandatory that high school graduates first complete a period of manual labor before entering a university. An extreme case of this new policy is seen in the enrollment of highly gifted teenagers, ages 14–16, in the University of Science and Technology.* The emphasis on the acquisition of expertise is also seen in the establishment of graduate programs both in universities and at research institutes of the CAS.
8. Standard textbooks are being compiled for the entire country.
9. Ties with international science and technology are being strengthened. This contrasts again with the extreme interpretations of the doctrine of self-reliance of the recent past. It means not only importing more foreign technology, but also increasing the number and quality of academic exchanges, sending more Chinese students and scientists abroad for extended periods of study, and having more active Chinese participation in international scientific meetings. Of particular importance has been the first official U.S. Government Science and Technology mission to visit China on July 5, 1978, led by Dr. Frank Press, the President's Science Advisor.
10. Scientists and engineers are being assured that at least five-sixths of their time will be spent on professional work. This assurance is intended to counter the practice of the recent past when professional personnel were required to have extensive involvement in political meetings.

* See, *Peking Review*, **April 14, 1978 15.**

11. Funds for science and education are being increased. Unfortunately, we know little about the amounts involved. One can, however, point to likely areas of budgetary expansion. These include:
    (a) *Salaries.* Promotions and salary increases had been frozen for over 10 years. A large number of promotions to higher ranks, which carry salary increases with them, are now being given. In addition, the total number of scientists and engineers is to increase rapidly during the next eight years.
    (b) *Capital construction.* Many of the facilities we saw were inadequate to meet the ambitious program China is undertaking. Expansion and improvement of existing facilities is called for. In addition, a number of new institutes have been established recently, and Hua Kuo-feng, Teng Hsiao-p'ing, and Fang Yi have all spoken of the establishment of major new comprehensive research centers.
    (c) *Annual research budgets.* These too can be expected to increase, particularly research budgets for institutions of higher education.
12. Efforts at popularizing scientific knowledge will be expanded.

Taken together with the new emphasis on the importance of basic research, these policy measures represent a major redirection of scientific life and a major change in the environment for R&D in China. The delegation saw evidence that many of these points had already been implemented.

As noted, the new directions in science are part of a larger comprehensive modernization program. Hence, before discussing the details of current science policy and plans, it will be helpful to review the broad outlines of the larger programs for economic development. A useful discussion of this topic is found in Hua Kuo-feng's report on the work of the government given at the Fifth National People's Congress on February 26, 1978.* Hua's report contains the outlines of an ambitious 10-year economic development plan. The major sections of the plan are as follows:

1. *Agriculture.* Strenuous efforts are to be made to double or even triple grain production within the next eight years in the 12 large commodity grain bases and on all state

* *Peking Review* **March 10, 1978 10,** 18ff.

farms. Grain deficient, low-yield areas should attain self-sufficiency in 2–3 years. In addition, efforts will be made to increase cultivated acreage by land reclamation, soil improvement, and water control. Although much of this work will be locally initiated and supported, the state will continue major water conservation projects on the Yellow, Yangtze, Huai, Haiho, Liaoho, and Pearl Rivers.

Efforts will be made to perfect the system of agro-scientific research and agro-technical popularization, to ensure linkages between advanced research and grass roots problems. Targets for progress in the agro-technical area are seed improvement, farming methods, fertilizers, mechanization, and insecticides. Various sources of fertilizer should be "extensively explored, making a big effort to develop organic fertilizer, and making proper use of chemical fertilizer." At least 85% of "all major processes of farmwork" should be mechanized within 10 years. Both fertilizer and insecticides should be made better, cheaper, and designed to meet specific needs. The target of all these measures is to increase the value of agricultural output by 4%–5% per year between 1978 and 1985.

2. *Industry.* Increases in the value of industrial output are targeted at 10% per year during the same period. According to Hua, the 10-year plan calls for

> . . . the growth of light industry, which should turn out an abundance of first-rate, attractive, and reasonably priced goods with a considerable increase in per capita consumption. Construction of an advanced heavy industry is envisaged, with the metallurgical, fuel, power, and machine-building industries to be further developed through the adoption of new techniques, with iron and steel, coal, crude oil and electricity in the world's front ranks in terms of output, and with much more developed petrochemical, electronics, and other new industries. We will build transport and communications and postal and telecommunications networks big enough to meet growing industrial and agricultural needs, with most of our locomotives electrified or dieselized and with road, inland water and air transport and ocean shipping very much expanded. With the completion of an independent and fairly comprehensive industrial complex and economic system for the whole country, we shall in the main have built up a regional economic system in each of the six major regions, that is, the southwest, the northwest, the central south, the east, the north, and northeast China, and turned our interior into a powerful, strategic rear base.

China's basic industries, such as the mining, steel, power, fuel, and transportation industries, are to receive particular state attention. Some 120 large-scale projects are to be undertaken including the building or completion of 10 iron and steel complexes, 9 nonferrous metal complexes, 8 coal mines, 10 oil and gas fields, 30 power stations, 6 new trunk railways, and 5 key harbors. The completion of these will result in China having "14 fairly strong and fairly rationally located industrial bases." Also singled out in Hua's report are the machine building and petrochemical industries. The latter is to contribute to substantially increasing ". . . the ratio of such petrochemically produced raw materials as chemical fibres and plastics to all raw materials used in light industry."

Hua's report also includes a section on science, education, and culture. Hua notes in particular the expected contributions of science and technology in ". . . rapidly transforming the weaker links in our economy, that is, fuel, electricity, raw and semi-finished materials industries, and transport and communications." Hua also gave a prelude to the national research priorities (discussed below) that were announced at the National Science Conference in March by stating,

> We must strive to develop new scientific techniques, set up nuclear power stations, launch different kinds of satellites, and step up research into laser theory and its application, attach importance to the research on genetic engineering and above all to research on integrated circuits and electronic computers and their widespread application. Full attention must be paid to theoretical research in the natural sciences, including such basic subjects as modern mathematics, high energy physics and molecular biology.

As a boost to scientific morale and to stimulate scientific and technological development, Hua Kuo-feng suggested in 1977 that a major National Science Conference be held in the spring of 1978. The Conference was to recognize and reward scientific achievement and to discuss new plans for science. It was preceded by a number of smaller, preparatory conferences during 1977. The national conference itself was held in March 1978, and Hua Kuo-feng, Teng Hsiao-p'ing, and Fang Yi all made addresses to it. The clearest statement of China's current policies and priorities was offered by Fang Yi in his report to the conference.*

In his report Fang notes the importance of having the plan for science and technology integrated with economic plans. The two must be organically combined, as with the 12-year plan for science

* "'Abridgement' of Fang Yi Report to National Science Conference," New China News Agency, March 28, 1978. In *FBIS*, March 29, 1978.

drafted in 1956. The latter is being treated as a model for the current efforts.

Fang then goes on to present the main features of the new "Draft Outline National Plan for the Development of Science and Technology, 1978–85." The broad objectives of the plan are

1. To reach "advanced world levels of the 1970's in a number of important branches of science and technology" in order to narrow ". . . the gap (with the advanced countries) to about ten years" and to lay ". . . a solid foundation for catching up with or surpassing advanced world levels in all branches in the following 15 years."
2. To "increase the number of professional research workers to 800,000."
3. To "build a number of up-to-date centers" for research.
4. To "complete a nationwide system of scientific and technological research."

The plan calls for comprehensive preparations for research in 27 "spheres," including natural resources, agriculture, industry, national defense, transport and communications, oceanography, environmental protection, medicine, finance and trade, culture, and education, ". . . in addition to the two major departments of basic and technical sciences." Within these 27 spheres, 108 items of research have been selected as "key projects."

Eight "comprehensive scientific and technical spheres, important new technologies and pace-setting disciplines that have a bearing on the overall situation . . ." were singled out for special prominence. These are:

1. *Agriculture.* Fang's discussion of agriculture follows Hua Kuo-feng's speech to the National People's Congress cited earlier, but is somewhat more detailed.

> We should implement in its entirety the Eight-Point Charter for Agriculture (soil, fertilizer, water conservancy, seeds, close planting, plant protection, field management and improved farm tools) and raise our level of scientific farming so as to bring about a big increase in agricultural output. We should study and evolve a farming system and cultivating techniques that will carry forward our tradition of intensive farming and at the same time suit mechanization, and manufacture farm machines and tools of high quality and efficiency. We will study science and technology for improving soil, controlling water, drastically changing the conditions of our farmland and turning it into crop fields that give stable and high yields. In order to improve as quickly as possible the low-yielding farm-

land that accounts for about one-third or more of the country's total, we must make major progress in improving alkaline, lateritic, clay and other kinds of poor soil, in preventing soil erosion and in combating sandstorms and drought. We will study projects for diverting water from the south to the north and relevant scientific and technical problems; study and develop new compound fertilizers and biological nitrogen fixation, methods of applying fertilizer scientifically and techniques for drainage and irrigation, cultivate new seed strains, develop new techniques in seed cultivation and improve the fine crop varieties in an all-round way so that they will give still higher yields, produce seeds of better quality and can better resist natural adversities. We should quickly find out new insecticides that are highly effective and are harmless to the environment, and develop techniques for simultaneous prevention and treatment of different kinds of plant diseases and pests.

We need to step up scientific and technological research in forestry, animal husbandry, sideline* production and fisheries and promote an all-round development of these branches. We should provide new tree seeds and techniques that will make the woods grow fast and yield more and better timber, develop multi-purpose utilization of forest resources and study techniques and measures for preventing and extinguishing forest fires; step up research on building pasturelands, improving breeds of animals and poultry, mechanizing the process of animal husbandry, increasing water life production, fish breeding, marine fishing and processing so as to make our contribution to improving the ingredients of the people's diet.

2. *Energy.*

We have our own inventions in the science and technology of the oil industry, and in some fields we have caught up with or surpassed advanced levels in other countries. We must continue our efforts to catch up with and surpass advanced world levels in an all-round way. We should study the rules and characteristics of the genesis and distribution of the oil and gas in the principal sedimentary regions, develop the theories of petroleum geology and extend oil and gas exploration to wider areas; study new processes, techniques and equipment for exploration and exploitation and raise the standards of well drilling and the rate of oil and gas recovery; and actively develop crude oil processing techniques, use the resources rationally and contribute to the building of some ten more oilfields, each as big as Tach'ing.

* Nonagricultural rural production.

> China has extremely rich resources of coal, which will remain our chief source of energy for a fairly long time to come. In the next eight years, we should basically mechanize the key coal mines, achieve complex mechanization in some of them and proceed to automation. The small and medium sized coal mines should also raise their level of mechanization. Scientific and technical work in the coal industry should center around this task, with active research in basic theory, mining technology, technical equipment and safety measures. At the same time research should be carried out in the gasification, liquefaction and multi-purpose utilization of coal and new ways explored for the exploitation, transportation and utilization of different kinds of coal.
>
> We must push up the power industry as a pressing task. We should take as our chief research subjects the key technical problems in building large hydroelectric power stations and thermal power stations at pit mouths, large power grids and super-high-voltage power transmission lines. China has a great abundance of water power resources. We must concentrate our efforts on comprehensive research in the techniques involved in building huge dams and giant power generating units and in geology, hydrology, meteorology, reservoir-induced earthquakes (as received) and engineering protection which are closely linked with large-scale hydroelectric power projects.
>
> We should devote great efforts to exploring new sources of energy in order to change China's energy pattern gradually. Atomic power generation is developing rapidly in the world, and we should accelerate our scientific and technical research in this field and speed up the building of atomic power plants. We should also step up research in solar energy, geothermal energy, wind power, tide energy and controlled thermonuclear fusion, pay close attention to low-calorie fuels, such as bone coal, gangue and oil shale and marsh gas resources in the rural areas, and make full use of them where possible.

In addition, Fang noted the importance of exploring techniques for energy conservation.

3. *Materials.*

> Steel must be taken as the key link in industry. Great efforts must be made to grasp metallurgical science and technology. It is imperative to make a breakthrough in the new technology of intensified mining and solve the scientific and technological problems of beneficiating hematite so as to provide the iron and steel industry with large quantities of raw materials. We should speed up research work on the paragenetic deposits at Panchihhua, Paotow and Chinchuan where many closely associated metals have been formed, solve the major technical

problems in multi-purpose utilization, intensify research on the exploitation of copper and aluminium resources, make China one of the biggest producers of titanium and vanadium in the world and approach or reach advanced world levels in the techniques of refining copper, aluminum, nickel, cobalt and rare-earth metals. We should master modern metallurgical technology quickly, increase varieties and improve quality; study and grasp the rules governing the formation of high-grade iron ore deposits and the methods of locating them; establish a system of ferrous and non-ferrous materials and extend it in the light of the characteristics of our resources.

We should make full use of our rich natural resources and industrial dregs and increase at high speed the production of cement and new types of building materials which are light and of high strength and serve a variety of purposes; step up research in the technology of mining and dressing non-metal ores and in the processing techniques; lay stress on research in the technique of organic synthesis with petroleum, natural gas and coal as the chief raw materials, step up our studies of catalysts and develop the technology of direct synthesis; renovate the techniques of making plastics, synthetic rubber and synthetic fiber and raise the level of equipment and automation in the petrochemical industry. We must solve the key scientific and technical problems in producing special purpose materials, structural materials and compound materials necessary for our national defense industry and new technology and evolve new materials characteristic of China's resources.

4. *Computer science and technology.* Fang notes that "... the electronic computer is developing in the following directions: giant computers, microcomputers, computer networks and intelligence simulation. The scientific and technical level, scope of production and extent of application of computers has become a conspicuous hallmark of the level of modernization of a country." Accordingly, China must rapidly solve problems relating to the industrial production of large-scale integrated circuits, work on the technology of ultra-large-scale integrated circuits, peripheral equipment, software, applied mathematics, and the problem of computer applications. Efforts will also be made to popularize microcomputers and establish a number of computer networks and data bases. By 1985, China hopes to have "a comparatively advanced force in research in computer science and build a fair-sized modern computer industry." Key industries will use computers for process control and management.

5. *Lasers.* Fang notes the applications of lasers in material processing, precision measurement, remote ranging, holography, telecommunications, medicine and seed breeding, and the potential application in isotope separation, catalysis, information processing, and controlled fusion. For the future, Fang stated,

> We will study and develop laser physics, laser spectroscopy and non-linear optics in the next three years. We should solve a series of scientific and technical problems in optical communications, raise the level of routine lasers quickly and intensify our studies of detectors. We expect to make discoveries and creations in the next eight years in exploring new types of laser devices, developing new wave-lengths of lasers and studying new mechanisms of laser generation, making contributions in the application of lasers to studying the structure of matter. We plan to build experimental lines of optical communications and achieve big progress in studying such important projects of laser applications as separation of isotopes and laser-induced nuclear fusion. Laser technology should be popularized in all departments of the national economy and national defense.

6. *Space.* Again Fang summarized the scientific and technological progress achieved in other countries as a result of the development of space technology. China's work in this field will include space science, satellites and ground facilities for remote sensing and other applications, the building of modern space centers, and the development of launch vehicles and skylabs.
7. *High Energy Physics.* In 1972 the Academy of Science established a new institute for high-energy physics. The plans for high-energy physics, according to Fang, are as follows:

> We expect to build a modern high energy physics experimental base in ten years, completing a proton accelerator with a capacity of 30,000 million to 50,000 million electron volts in the first five years and a giant one with a still larger capacity in the second five years. Completion of this base will greatly narrow the gap between our high energy accelerators and advanced world levels and will stimulate the development of many branches of science and industrial technology.
>
> We should from now on set about the task in real earnest and make full preparations for experiments in high energy physics, with particular stress on studying and manufacturing detectors and training laboratory workers. We should step up research in the theory of high

> energy physics and cosmic rays, consciously promote the interpenetration of high energy physics and the neighbouring disciplines, actively carry out research in the application of accelerator technology to industry, agriculture, medicine and other spheres, and pay attention to the exploration of subjects which promise important prospects of application.

A more recent report states that work on the 30–50 Bev. proton synchrotron has started in Peking. The accelerator is to be completed by 1982. A machine of higher energy is to be completed by 1987.*

8. *Genetic Engineering.* Fang says of this field that

> It is possible for genetic engineering, an outgrowth of molecular biology, to splice and transfer genetic substance at the molecular level and create new biological species to meet the needs of humanity. Genetic engineering provides an effective means of experiment for such basic studies concerning higher organisms as cell differentiation, growth and development and formation of tumours. It is likely to open new vistas for momentous changes in agriculture, industry, medicine and certain other fields of production.
>
> Genetic engineering is a new branch of study which appeared in the 1970's. Fast developing and highly explorative, it deals with a wide range of disciplines and technologies, yet our country has only a rather weak foundation in this respect. Therefore, we must in the next three years strengthen organization and coordination and step up the tempo of building and improving the related laboratories and conduct basic studies in genetic engineering. In the next eight years, we should combine them with studies in molecular biology, molecular genetics and cell biology and achieve fairly big progress. We should study the use of the new technology of genetic engineering in the pharmaceutical industry and explore new feasible ways to treat certain difficult and baffling diseases and evolve new high-yield crop varieties capable of fixing nitrogen.

Much of the remainder of Fang's report deals with general policies reiterating points made during the months preceding the Conference. These are dealt with under the following headings:

1. *Consolidate the scientific research institutions and build up a scientific and technological research system.*

Here Fang is addressing issues of leadership and organization. Of particular interest is (a) the restoration of the importance of having professionals in leadership positions; (b) the stress on the role of the

* *Peking Review* **June 9, 1978 23,** pp. 3–4.

authorities at provincial, municipal, and other subnational jurisdictions to promote research activities at those levels; and (c) the role of the Communist Party. The last point is of some importance since the active cooperation of Party committees throughout the country is vital to the achievement of the policy objectives. Much effort was made during 1977 and early 1978 by the central authorities to prepare subnational authorities for the tasks ahead.

In addition, the precise role of the Party at the institute level is of interest, because officially institute leadership is defined as "the system of institute directors assuming responsibility under the leadership of the Party committees." The nature of this Party leadership was spelled out somewhat by Teng Hsiao-p'ing in his speech to the Science Conference.*

Teng noted that the leadership given by Party committees is primarily political leadership. The Party is responsible for the correct political orientation of the work of the institute, and for seeing that current policies and principles of the Party are being enforced. The Party committee is also to be active in planning, in seeing that plans are carried out, and in facilitating formulation of new plans within the institute. We were told that one of the functions of the Party is, with the institute director, to resolve disagreements between the planning section of the institute and the academic committee, should such disagreements arise.

Teng also specified that the Party committee should act as a facilitator of research work by ensuring that facilities, supplies, supporting services, and general working conditions are adequate for the task.

As Teng himself put it to the Conference, "I am willing to be the director of the logistics department at your service, and do this work well together with the leading comrades of Party committees at various levels."

The dramatically changed roles of scientists and Party committees is further highlighted in Teng's speech as follows:

> We should give the director and the deputy directors of research institutes a free hand in the work of science and technology according to their division of labor. Party committees should back up the work of all Party and non-Party experts in administrative positions and try to bring out all their capacities so that they really have powers and responsibilities commensurate with their positions. These experts are also cadres of the Party and the state. We must never look askance at them. Party committees should get acquainted with their work and examine it but should not attempt to supplant them.

* *Peking Review* **March 24, 1978 12.**

And to drive home to the Party its proper role, Teng stated,

> The basic task of scientific research institutes is to produce scientific results and train competent people. They must show more scientific and technical achievements of high quality and train scientific and technical personnel who are both red and expert. The main criterion for judging the work of the Party committee of a scientific research institute should be the successful fulfillment of this basic task. Only when this is well done has the Party committee really done its duty to consolidate the dictatorship of the proletariat and build socialism. Otherwise, putting politics in command will remain merely empty talk.

Fang's report continues:

2. *Open broad avenues to able people and recruit them without overstressing qualifications.*

This section of Fang's report urges a flexible approach to building an expanded cadre of R&D personnel. Although the recruitment system is becoming somewhat more formalized through the institution of examinations for universities and postgraduate programs, Fang insists that unconventional means be maintained for identifying talent from unexpected quarters. Fang specifies possible candidates for recruitment as participants in science contests, self-educated readers of journals, and inventors and innovators from agriculture and industry. In addition, exceptionally able students can be accelerated, as in the case of the teenagers of the University of Science and Technology cited above.

3. *Institute regulations for training, testing, promoting, and rewarding scientific and technical personnel.*

Included in this section are provisions for titles and promotions and for the selection of research workers for advanced study abroad. Outstanding achievements should be rewarded in various ways: "Moral encouragement should be the main form but there should also be proper material rewards."

4. *Uphold the policy of letting a hundred schools of thought contend.*

5. *Learn advanced science and technology from other countries and increase international academic exchanges.*

6. *Ensure adequate work hours for scientific research.*

7. *Strive to modernize laboratory facilities and information and library work.*

In this section Fang raises a number of interesting points about the infrastructure for Chinese science. They are of interest in connection with observations made by the delegation elsewhere in this report. Relevant excerpts of Fang's speech include the following:

> In the next eight years we should build a number of modern experimental installations and centers. We should give a high priority to refitting the existing laboratories so as to modernize them as quickly as possible.
>
> . . .
>
> Emergency measures must be taken to push forward the designing and production of instruments and equipment. Efforts must be made to expand, renovate and build a number of factories specializing in scientific instruments and chemical reagents. Scientific research institutions, universities and colleges should pay great attention to new principles, new techniques and new products in their research on instruments and equipment and where necessary expand the capacity for processing, trial-manufacturing and production.
>
> . . .
>
> It is essential to strengthen the management of the designing, production, distribution and use of scientific instruments and bring them under an overall national plan. Costly large precision instruments should be used jointly by the units requiring them so that they are fully utilized. All large modern experimental centers should be open to scientific and technical personnel from organizations related to them and teachers and students from universities and colleges who come to conduct experiments and research so that these centers will gradually become research complexes.
>
> . . .
>
> With the development of science and technology, the number of scientific papers and data is increasing tremendously. Several million scientific papers are published in the world every year. If we should fail to keep abreast of the developments, trends and levels of achievement in science and technology the world over, and waste our valuable man and material power in following the beaten track and making detours that others have made, it would be out of the question for us to reach, catch up with and surpass advanced world levels in science and technology.
>
> . . .
>
> It is essential to modernize scientific and technical information work and equip information institutions with modern facilities in the shortest possible time. In the next eight years we will set up a number of documentation retrieval centers and data bases and build a preliminary nation-wide computer network of scientific and technical information and documentation retrieval centers. We should also strengthen the publication of scientific and technological material.

8. *Close cooperation with an appropriate division of labor.*

The main institutional arrangements of research are spelled out as follows:

> The Chinese Academy of Sciences is the overall center for research in natural science throughout the country. Its main task is to study and develop new theories and techniques and to solve major scientific and technical problems involving many fields of economic construction, in cooperation with the departments concerned. It should lay stress on basic theoretical research and aim at raising standards. The institutions of higher learning serve as both educational and research centers; they are an important force in scientific research, covering both the basic and the applied sciences. Research institutions of the various departments and localities should devote themselves mainly to the applied sciences, but they should also undertake appropriate research in basic science. The above institutions and the nonprofessionals engaged in scientific experiment should work in close cooperation with an appropriate division of labor.

9. *Speed up popularization and application of scientific and technical achievements and new technologies.*

In this section Fang discusses the key task of linking research with production and identifies in particular the importance of pilot factories and workshops for trial production of new products or uses of new processes. He also shows an awareness of the dynamics of the innovation chain and of the problems of technical conservatism among managers in a planned economy, noting that "The standards by which production departments are examined should include the application of such (scientific and technological) achievements, and the innovations made in technology. We should actively support their efforts to apply new techniques and improve work processes by providing them with the necessary materials and funds."

10. *Make painstaking efforts to popularize science.*

Fang's speech is the most comprehensive public science policy document produced in China in almost 20 years. In addition to the speech, the Delegation also had access to Chinese scientific policy thinking in discussions with our Chinese hosts. In particular, a subgroup of the Delegation had a three-hour evening meeting to discuss scientific policy questions with Ch'ien San-ch'iang, a Vice-President of CAS, and with Li Su, a Deputy Secretary-General of CAS. At the meeting Ch'ien noted that although basic research was neglected in the past, current policy was supportive. He acknowledged that part of the credit for turning China's attention to basic research belongs to American and other foreign scientists who criticized China's neglect of basic research. This criticism was highly valued.

We discussed the rationale behind the eight priority areas. The first three—agriculture, materials, and energy—are intended to subsume a broad range of research of direct importance to China's modernization goals. The next three—computers, lasers, and space—were seen as "levers" with important implications for a broad range of scientific pursuits and of actual and potential applications. Genetic engineering and high-energy physics were seen as key areas of basic research dealing with the fundamental nature of living organisms and with the structure of matter and were thus viewed as having potential implications for other fields of science.

The inclusion of high-energy physics was discussed in some detail in response to skeptical questioning by members of the Delegation. Ch'ien seemed quite familiar with the arguments against a major commitment to high-energy physics, and it seems that these arguments had also been made in China. Ch'ien's position, however, was that one could not predict whether useful discoveries will be made in such an area of research. China wants to be in a position to make useful discoveries in this field if they are to be made. It was clear that the highest political authorities were behind the development of high-energy physics. Ch'ien pointed out that Mao Tse-tung had long been interested in questions about the structure of matter, and Teng Hsiao-p'ing was quoted as saying that it is imperative for a socialist country to have a research capability in this field.

Ch'ien was also asked why medicine and public health were not included in the eight priority areas. He explained that this question was discussed at great length during the preparation of the plan and that medicine came very close to being included. Li Su added that even though medical research is not one of the eight, it should not be interpreted to mean that the government does not consider it important.

Ch'ien also provided some insight into the processes of policy-making. He acknowledged that there were competing interests and constituencies in the formulation of plans and much "struggle" and debate at the national planning conference. He also remarked that the planning process may vary from field to field. In chemistry, plans were developed by attempting to reconcile three major considerations: (1) the relevance of areas of chemical research to the needs of agriculture, industry, and national defense; (2) the relevance of areas of chemical research to other areas of science; and (3) the important areas of research internal to chemistry itself. Ch'ien noted that the reconciliation of these areas has been imperfect and that research policy in chemistry—and, by extension, research policy in general—has had a zigzag course and that many mistakes have been made in the past. One of the goals at present is to minimize the deviations from the correct course and to keep science on a consistent and relatively straight path of development. Ch'ien stated that he expected that the new leadership provided by the reestablished Science and Technology Commission will greatly improve policy consistency. He also stated that the STC has an important role in coordinating the different research sectors.

4. CHEMISTRY AND CHEMICAL ENGINEERING AS ELEMENTS OF SCIENCE EDUCATION IN CHINA

A. Legacy of the Cultural Revolution

The Cultural Revolution and its aftermath, which spanned the decade 1966–76, greatly disrupted the university system in China. Teachers at all levels were included with other intellectuals as a class considered to be in need of ideological remolding. Under the influence of the Gang of Four in the early 1970s, many were considered enemies of the state. During this period the undergraduate college program was reduced to three years and even during the three years of undergraduate training, much time was spent on indoctrination in the philosophy of the Cultural Revolution. Students and teachers alike spent significant time out of the academic year working on farms. In addition, many intellectuals were actually persecuted. There were numerous reports about students trained during this period who had inadequate academic preparation for a career in science even at the level expected from someone with a reasonable undergraduate training in chemistry or chemical engineering.

The senior people active in teaching and research, both at the university and institute level in China, include a group who are over the age of 55 and were predominantly educated at the Ph.D. level in the United States. It was a tradition to send outstanding students from China to the United States, Europe, and Japan for both undergraduate and graduate education prior to 1950. Inasmuch as there is no formal retirement age for university faculty or institute professors, this group has come to dominate the leadership positions in the major universities and research institutes. A second group consists of scientists in the age range 45–55, some of whom were educated at the graduate level in the Soviet Union between 1950 and 1960, with the rest graduating from pre-Liberation universities either in the late 1940s or the early 1950s. The generation of scholars between the ages of 30 and 45, who had only limited access to training in China, represent a nearly lost cohort. This has resulted in a serious shortage of trained personnel, which will impact the ability of the country to train the next generation of scholars and to advance their levels of basic research in the next two decades.

B. Chemistry

The educational system in China starts with primary school at age six. Children are educated for five years in primary schools. This is followed by three years in a middle school, and two years in a senior school. Students who intend to enter the university in a program of science will typically study chemistry and physics each for two years in middle school or senior school.

Undergraduate programs

The projected curriculum for training in chemistry at the undergraduate level starting in 1978 is as follows:

First year	general chemistry, mathematics, physics, English, and political science (Marxism and Maoism)
Second year	analytical chemistry, organic chemistry, physics, and English
Third year	physical chemistry, structural chemistry, industrial chemistry, and organic chemistry
Fourth year	specialized courses with particular emphasis on catalysis and structural chemistry

The development of a meaningful curriculum at the levels we see in the West will take some time. Most university faculties in China are depleted, the teaching laboratories are antiquated, and there is a deficiency of modern textbooks in the Chinese language. We visited the following universities: Peking University, Tsinghua University (Peking), Kirin University (Ch'angch'un), Futan University (Shanghai), Chekiang University (Hangchow), and Northwest University (Sian). Of these, only Kirin University and Futan University have reasonable laboratory facilities and academic programs at a level comparable with those in the West. Peking University, Futan University (Shanghai), and Kirin University (Ch'angch'un) are among those included in the list of 88 universities identified as centers of excellence in higher education.

Graduate Programs

In the fall of 1978 students were admitted to graduate programs in chemistry among other disciplines. The method of selection of students for admission to graduate programs is detailed in Chapter 4. Both university departments and institutes of the Academy of Sciences accepted graduate students in the fall of 1978. Primary responsibility for basic research in science has been designated to the Institutes of the Academy of Sciences under the National Science Plan. The universities have a secondary role in the conduct of basic research. The institutes of the Academy of Sciences suffered much less during the Cultural Revolution than the universities. They are therefore better prepared to start students in a research program.

The establishment of formal graduate programs in the institutes of the Academy of Sciences will be an interesting experiment. For example, in the Max Planck Institutes in West Germany, the graduate student is formally enrolled at an affiliated university. The student carries out research with a professor who has an appointment at the Max Planck Institute as well as a formal connection with the university. In general, the Max Planck Institute is located on the same grounds as the university. Academic credentials are carefully monitored by the university faculty. A similar collaborative arrangement exists with respect to graduate students in the Soviet Union. However, in the Soviet Union almost all graduate research is done at institutes of the Academy of Sciences. The universities in the Soviet Union are primarily teaching institutions and research is usually not at the same levels as in the institutes of the Academy of Sciences. However, in the Soviet Union the student does take formal course work and receives a degree from the university where the professor also holds a faculty position.

The program envisaged in China will have complete independent graduate programs at the institutes of the Academy of Sciences. Students will receive graduate instruction, will take formal examinations, and will carry out their research at the institutes. This educational experience will be narrow since it will deny a graduate student the richness that comes from interaction with graduate students in other subdisciplines of the field or in completely different fields. It will be interesting to watch the development of this program. It will also be interesting to see how the parallel programs in universities compare with those at the research institutes.

C. Chemical Engineering

There is a wide variety of arrangements for chemical engineering departments in China. It is generally agreed that the five best departments are Shanghai Chemical Engineering College in Shanghai, Chekiang University in Hangchow, Tientsin University in Tientsin, Tsinghua University in Peking, and the Talien Engineering College. Other departments often mentioned as very good include Ch'engtu Engineering College in Ch'engtu at Szechuan, Kuangtung Chemical Engineering Institute in Canton, Northwest University in Sian, and the Peking Chemical Engineering Institute. There are also two specialized petroleum educational institutes at the oil fields of Tach'ing and Shengli.

The Shanghai Chemical Engineering Institute is a freestanding school for chemical engineering after the Russian model. It was established in 1952 by combining the chemical engineering departments of five different universities in East China. Chekiang University has colleges of both science and engineering, while Tientsin University, Tsinghua, and Talien have programs of engineering only. Northwest University has arts, science, and engineering, although the only department in the engineering college is chemical engineering. The special schools in petroleum technology at Tach'ing and Shengli comprise the entire spectrum of education necessary for the petroleum industry, including geology, oil drilling, oil recovery, refining, and transportation by pipeline.

The freestanding Shanghai Institute for Chemical Engineering was established under Russian influence, which has been much diminished in recent years. Together with a chemical engineering institute, there might also be freestanding technological institutes for mechanical engineering, civil engineering, construction, architecture, electrical engineering, and communications in a given city. Since these freestanding institutes are often very far from each other geographically, each must have a department of mathematics, physics, chemistry, as well as a library and computing facilities. These freestanding institutions obviously represent a great deal of duplication of effort and reduce the chances of cooperation. Just to give an idea of the duplication that exists, the city of Sian has the following universities of higher education: Agriculture, Highways, Foreign Languages, Art, Law, Electrical Communications, Mechanical Engineering, Athletics, Transportation, Medicine, Construction and Metallurgy, Mining and Coal, the Normal Teacher's College, the Northwest Industrial University, and the College of Light Industry.

Despite the unpopularity of the Russians, once these freestanding technological institutions were established, they were very difficult to abolish or to recombine into institutions of a more economic size. There is much talk that an ideal model is the Massachusetts Institute of Technology, where colleges of science and engineering are together.

The ratio of students to faculty at a Chinese university is very different from that usually encountered in the United States. For instance, in the Shanghai Chemical Engineering Institute, there is a teaching staff of 990 of which 68 are professors and associate professors and the rest are lecturers and instructors. All these are permanent, full-time positions. They have a student body of 2600. Therefore, the ratio of student to teacher is about 2.7. This is more or less typical of all the Chinese universities we visited. Many of the instructors do not give lectures, but help the professors by discussing homework problems and exams with students and by assisting in the teaching laboratories. Most of these teaching universities have had a three-year program, but now all are switching to a four-year program. The larger departments, such as the Shanghai Chemical Engineering Institute, are divided into sections such as petroleum chemical engineering, organic chemical engineering, inorganic chemical engineering, chemical engineering machinery, and fundamental training. Since this is a freestanding chemical engineering college, fundamental training includes topics such as mathematics, chemistry, physics, and languages.

During the period of the Cultural Revolution and the Gang of Four, entrance examinations were abolished and so were course examinations. The result was a drastic decline in the quality and preparation of the students. Entrance exams have been reintroduced with emphasis on theoretical and fundamental preparation. There is a national plan that will be coordinated by the Ministry of Education and the Ministry of Chemical Industry to revise the curriculum and to write a new set of textbooks. There have been few textbooks written in the past 15 years, and the older ones are very much out of date. There were essentially no graduate students during the period of the Cultural Revolution and the Gang of Four, but graduate students are now beginning to be reintroduced in the chemical engineering departments. Graduate students are expected to study for a period of three years that includes both courses and thesis; they do not receive a degree such as Ph.D. but a certificate for having participated. Most faculty members did not want to discuss their current curricula because they felt that they were in a state of flux. However, they were very interested in discussing the curriculum in chemical engineering programs at leading American universities.

The current leadership of the Departments of Chemical Engineering is mostly in the hands of U.S. graduates who returned to China during the 40s and early 50s. They have fond memories of their time in the United States and are anxious to resume contact. After this age group, there was a 10-year period during which most of the chemical engineering leaders were trained in the Soviet Union, followed by another period where few students went abroad for studies. This is also reflected in the textbooks in the libraries. The older textbooks are essentially from the United States, followed by more recent books from the Soviet Union, West Germany, and Japan. The Chinese are very anxious to find out what are the best-selling textbooks in the United States and how to receive copies of them. The pattern is to choose the best-selling textbooks in the United States and to translate or reedit them for the Chinese students. The task is coordinated by a central body of the Ministry of Education and the Ministry of Chemical Industries, which assigns the task of writing textbooks in different areas to different universities.

A typical curriculum, such as that found in Chekiang University, consists of a series of required courses and no electives. The required courses include 2 terms of mathematics, 2 terms of physics plus laboratory, 2 terms of inorganic chemistry plus lab, 2 terms of organic chemistry plus lab, 2 terms of analytical chemistry plus lab, and 2 terms of physical chemistry plus lab. In chemical engineering: 1 course in thermodynamics, 2 in transport phenomena, 1 in reaction engineering, 1 each in polymer chemistry, physics, and technology, 1 in petrochemical engineering, which is mostly applied thermodynamics, 1 in equipment, 1 in design principles, and 2 in unit operations laboratory. The curriculum also includes politics (Marx and Mao), 6 semesters of English, physical education, mechanical and electrical engineering, and drawing. A particularly interesting section in chemical engineering colleges is "chemical machinery," which is seldom found in the United States today. Most departments have very simple computers for research and none for teaching.

The chemical engineering laboratories are quite well equipped and adequate for undergraduate instruction. In fact, they are much better appointed than many of the laboratories in the United States today. It is quite common to have large and well-designed unit operations equipment that is in good working order. Many of the laboratory units have good instruments but little or no computation facilities. These same labs would be inadequate for serious graduate research.

D. The Selection of Students for Undergraduate and Postgraduate Programs

With the fall of the Gang of Four and the end of the Cultural Revolution, China's universities are entering a new phase. In December 1977 the first national college entrance examination was administered in examination centers all over the country. In Peking alone there were 95 such examination centers. The tests ran for two-and-a-half days and are part of a major reform of the college admissions system introduced in 1977.

During the most extreme period of the Cultural Revolution in which there was a general assault on intellectuals, the idea that students should be selected on the basis of their skills was derided, and a student who handed in a blank examination paper was acclaimed a hero. The criteria during this 10-year period were varied, but, of course, the college programs to which students were admitted during the Cultural Revolution were so disrupted that it was perhaps unimportant how the students were chosen. Now that China is making a serious effort to promote science and technology and, in general, to promote excellence in trained personnel, the new admissions system is extremely demanding and selective. This year is unusual since there is a large backlog of students who could not enter previously disrupted universities, and as many as five million people applied for admission to college. These include not only students graduating from secondary schools but also others who have been serving as workers for a year or more but who now want to receive a university education.

Students may list their first three choices of universities and the specific program in each of these universities that they will pursue. Admission by the university is based principally on the test scores, and the difficulty of admission depends not only on the standards of a particular university but also on the standards of the particular program that the student has specified in each university. A student might list the chemistry program in Futan University (in Shanghai) as a first choice, but a second choice might be the biochemistry program in a completely different university. The number of people to be admitted to each of these programs is set in line with national goals, and the students, along with their parents and advisors, engage in an elaborate guessing game to try to anticipate the popularity of various of these fields and thus the probability of admission to particular programs. Students admitted to a program must stay with it, although students who wants to go on to work in chemistry can apparently do this successfully even if their area of undergraduate preparation has been in one of the allied fields like biochemistry or branches of applied chemistry. The estimates we received from a variety of sources are that perhaps as few as 5% of the applicants this year have actually been admitted to any university. Those who failed to be admitted go to work in a factory or farm, meanwhile preparing for the examinations and the application procedures in future years until they are successful or become discouraged.

Most universities are just starting graduate programs after the 10-year hiatus of the Cultural Revolution. New graduate classes were admitted for the first time in the Fall of 1978 after the procedure for this admission had been worked out. National examinations are prepared in Peking and distributed along with answer keys. The individual chemistry departments in universities or in research institutes (many of which are also admitting graduate students, although they do not generally participate in undergraduate teaching) then look over the pool of applicants to their program and make a preliminary choice of perhaps twice the number they actually intend to take. The finalists are then brought to the university or research institute at the expense of the institution and undergo final screening by means of a locally prepared examination. Typically between 10 and 30 students were admitted in chemistry by each of the major institutions in the fall of 1978, although many institutions indicated that this was simply a way to get the programs started and that the numbers might increase in future years. As these procedures were being applied for the first time in 1978, it is possible they will be changed with experience. The graduate program does not actually lead to the awarding of a degree but to a diploma certifying completion of the program.

E. Student Placement

Students who finish their undergraduate work may try to go on to graduate school, but for many of them undergraduate work completes their formal training, and they must enter the ranks of the employed. There is apparently no unemployment problem, and everyone can find some kind of job. During the Cultural Revolution China suffered extensively from underuse of trained manpower, since many scientists were assigned to menial labor. However, that period seems to be over, and China is using trained manpower in areas for which the skills are relevant.

At most research institutes and plants we visited, there were complaints about the serious shortage of manpower, and managers said they were lucky to get one engineer out of three requests. Placement seems to be under the control of the State Planning Commission and the Ministry of Education in Peking, with inputs from the Ministries of Petroleum and of the Chemical Industry. Requests for manpower are collected by the Ministries in Peking from the various organizations that require personnel; the ministries then decide how many engineers each organization gets and who they shall get. The ministries work mostly from dossiers and letters of recommendation from the university committees. There are some organizations that receive significant help from the local city bureaus of chemical industry or textiles in obtaining the needed manpower for their operations.

The universities themselves play a major role in job assignments. The university administration receives a notice of jobs that are available and is asked to recommend specific candidates for these positions. Frequently a representative of the agency that requires technical manpower will come to the universities to look over the records of the candidates, but there seem to be no personal interviews of candidates by employers before job selections are made.

Students do not openly discuss their job preferences. A few of them explain that they have taken entrance exams for the graduate schools and are waiting for the results. The rest would only say that they would go wherever the nation needs them—they have no preferences. Many of the current group of undergraduate students are 24 to 25, which seems old by U.S. standards. However, they have often spent a number of years on farms or in factories. When asked if they would be interested if there were a chance for them to study at a graduate school in the United States, one student answered that his class was too poorly prepared and that perhaps the next few classes would be better prepared and be more deserving of the opportunity.

The apparent willingness to take any job assigned may be bolstered by the fact that salaries are remarkably uniform. In Peking essentially all jobs that come after the conclusion of the undergraduate program pay a starting salary of 56 yuan per month (a little less than $40). Starting salaries are slightly different in other locations, presumably in recognition of variations in costs and attractiveness, but they are all fundamentally of this magnitude. The most desirable jobs seem to be in the research institutes or in the universities, while the least desirable are in factories. Much of the kind of research that is performed in the research laboratories of industry in our country is performed in China in ministry or academy research institutes, and jobs in the factories themselves tend to be rather routine in character. The government itself assigns priorities to the available jobs, so that the best candidates tend to be placed in the positions of greatest importance and desirability.

5. BASIC RESEARCH IN THE SUBDISCIPLINES OF CHEMISTRY AND CHEMICAL ENGINEERING

A. Organic Chemistry

During the Cultural Revolution, organic chemistry as a broadly defined field was considered to be somewhat irrelevant to the immediate needs of the country. It shared this distinction with physical chemistry, inorganic chemistry, and the other classically defined academic areas. In the universities, faculty research programs were brought essentially to a halt, and many of the faculty were dispersed to factories where they performed nonscientific work. The teaching laboratories were also disrupted and in many cases physically dismantled. Now research programs in the universities are just getting started again, and in most cases we heard only about plans for the future. New graduate students came in the summer of 1978 to reinstitute graduate programs, and there is still active discussion about the fields of research that will be pursued with these students. Thus, the universities are not yet a significant factor in Chinese research in organic chemistry, and even their training function is just getting reestablished.

By contrast, in a few institutes reasonable programs with applied emphasis were pursued even during the worst days of the Gang of Four. These include, for example, the Institute of Applied Chemistry in Kirin, the Institutes of Materia Medica in Peking and Shanghai, and the Institute of Organic Chemistry in Shanghai, which in fact has devoted much of its attention in recent years to applied aspects of the field.

There are three major centers of organic chemical research at this time in China with a few others of potentially significant quality.

The Institute of Organic Chemistry, Shanghai

The leading program from the point of view of breadth and depth is here. Research at this institute is divided into three general areas. One of them can broadly be classed as natural products. This field includes work on peptides, on insect hormones, on prostaglandins, on nucleic acids, and on steroids. There is also some work on the biological oxidation of hydrocarbons obtained from petroleum.

As an example of the work on polypeptides, a structural study is now underway on trichosanthin, a protein isolated from a plant that has been used as an abortifacient in Chinese folk medicine and that also has demonstrated anticancer activity. The compound has molecular weight 240,000 and its isolation utilizing gel electrophoresis and gel chromatography was followed by a sequence determination using a polypeptide sequencer manufactured in the United States. Crystals have been obtained, and an X-ray study is currently being done.

In the insect area, synthetic work is underway on preparing mimics for the silkworm juvenile hormone, and some work is also going on on the chemistry of ecdysone. A pheromone of the piny bollworm has been identified as

$$C_4H_9\underset{\uparrow}{CH{=}CH}{-}(CH_2)_2{-}\overset{H}{\overset{|}{C}}{=}\overset{H}{\overset{|}{C}}{-}(CH_2)_6CO_2CH_3$$

a 1:1 cis/trans mixture

on the basis of some degradative work done elsewhere and synthetic work done in this Institute. The insect requires a one-to-one mixture of the cis and trans isomers at the double bond indicated by the arrow in order that the compounds be effective. A new sesquiterpene derivative has been isolated from artemisia, and is called arteannuin.

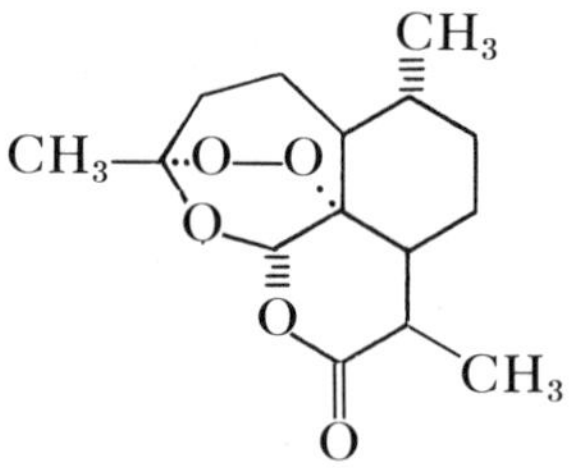

The unusual structure was demonstrated by a combination of proton NMR and mass spectroscopy, while the ^{13}C NMR spectrum had to be obtained at the Institute of Photography in Peking, currently the only place in China where such instrumentation is available. The X-ray crystal structure was performed at the Institute of Biophysics in Peking.

Work on prostaglandins was started in 1971. Some of these compounds are being prepared by biosynthesis, and others by total chemical synthesis. The routes involved are relatively sophisticated and involve the use of good current synthetic methodology.

In nucleic acids the most significant effort is a cooperative project along with the Institute of Biophysics in Peking to achieve the first synthesis of a transfer RNA, in particular of yeast alanine *t*-RNA. The people at the Institute of Biophysics are doing the enzymatic studies involving splicing of polynucleotides to produce larger pieces, while the Institute of Organic Chemistry is pursuing studies on the synthesis of smaller pieces. Elegant, modern techniques are used, including the application of ^{32}P with enzymatic ligase reactions in the synthesis of dodeca- and hexadecanucleotides. This project also has active collaboration with the Shanghai Institute of Experimental Biology and the Shanghai Institute of Biochemistry. It seems to represent in the nucleic acid area an attempt to repeat the tour de force of the insulin synthesis achieved before the Cultural Revolution.

In the steroid area there is considerable concern with producing needed medicinal steroids, including popular birth control agents, from available materials. Some work is being done on total synthesis using essentially a variant of the Torgov scheme, and in addition transformations starting from the available diosgenin and the increasingly available hecogenin and tigogenin are also underway.

The second general area of work in the institute is classed as organoelement chemistry, in line with the Russian description of this field. This includes not only organometallic chemistry but also synthesis and characterization of organofluorine, organoboron, and organophosphorus compounds, as well as those with tin and arsenic. Several programs are currently underway in this field. For example, studies of the reactivity and synthetic applications of arsenic ylides are being conducted. Arsenic ylides are more reactive than the corresponding phosphorus compounds, and thus lead to significant yields in cases where the analogous phosphorus reagents are inert. For example, in the reaction

$$(C_6H_5)_3As{=}CHCO_2Me + (C_6H_5)_2C{=}O \rightarrow (C_6H_5)_2C{=}CHCO_2Me + (C_6H_5)_3AsO$$

a 77% yield is obtained with the arsenic ylide, but the yield is 0% under the same conditions for the phosphorus analog. Work is also underway on polymers of fluorinated olefins and on the preparation of organophosphorus chelating reagents for selective extractions of metals. All the fluorinated intermediates must be prepared as well, and extensive studies have been pursued on mechanisms of fluorination of organic compounds. A class of stable fluorinated surfactants, whose structures are

$$CF_3(CF_2)_nO(CF_2)_mSO_3^-K^+$$

have been prepared as stable materials for applications in difficult conditions.

The third general area of work in the institute is physical organic chemistry. A few specific projects are underway. One involves detailed studies of liquid crystals with a view toward developing appropriate technology in this area. The effort here is illustrative of one of the styles of research we encountered. The Chinese are free to try to duplicate patented technology, since they do not subscribe to the international patent agreements, but they face the usual problem in trying to repeat work that is published only in the patent literature. For this reason one of their major efforts at this institute has been to try to explore the properties of the already existing patented liquid crystal materials that are used in display devices. In the course of the work a considerable expertise has been developed in the evaluation of these materials, and this is now being applied to the development of new liquid crystal substances and mixtures.

Another area being studied is free radical addition to unsymmetrical fluorinated olefins and the role of polar effects in determining such additions. This work was published in *Scientia Sinica,* **1977** *20,* 353. Other work includes a study of the thermal dimer of phenyltrifluoroethylene, whose structure had originally been misassigned by Russian workers but was corrected here and reported in *Acta Chemical Sinica* in 1976 more or less simultaneously with American work correcting the structure. Spectral and NMR studies are also being pursued on some small fluorinated compounds, and stereochemical and structural factors involved in fragmentation patterns and NMR coupling constants will be reported shortly in *Acta Chemical Sinica*. Another project in the general area of physical organic chemistry involves a detailed attempt to correlate the extraction properties of organophosphates with some of their other physical properties.

Instrumental support for the program of the Institute of Organic Chemistry is only modest. There is a pilot plant available for the preparation of large amounts of materials, and indeed many chemicals are prepared and instruments constructed in the Institute. The Institute machine shops have made high pressure liquid chromatography equipment shown in Fig. 8. Commercial instruments include standard infrared and UV machines, a mass spectrometer and GC/MS instrument, and a high resolution mass spectrometer. Only a relatively simple 60-MHz proton NMR machine is available here, with a 100-MHz cw NMR in the nearby Institute of Materia Medica. No ESR or Raman instrumentation is available. It is apparent that this Institute needs considerably better instrumentation, particularly in NMR.

The Institute of Organic Chemistry started training graduate students in the summer of 1978, and admitted approximately 15 new students for the first year's program.

Figure 8. High-pressure liquid chromatograph, Institute of Organic Chemistry, Shanghai.

Institute of Materia Medica, Peking

The areas involved in the program in this Institute include synthetic organic chemistry, phytochemistry, antibiotics, analytical chemistry, pharmacology, medicinal plants, and the cultivation of medicinal plants. Among the projects being pursued are some partial synthetic approaches to harringtonine, an anticancer compound that is in clinical use.

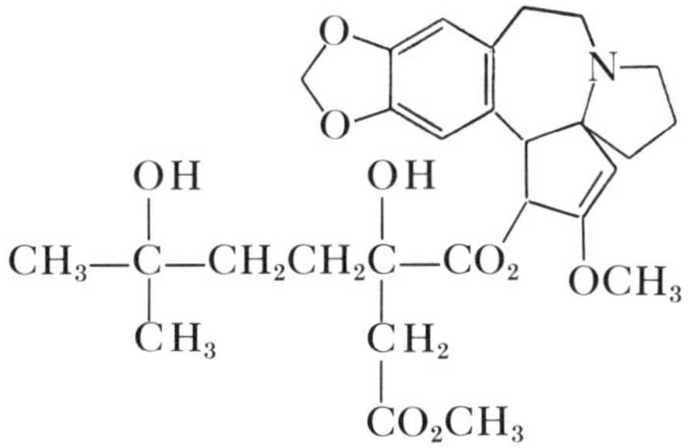

The dried fruit of Schizandra chinesis Baill is a widely used traditional Chinese drug which has been classified as one of the "superior medicines." It is used for the treatment of hepatitis. Work in the Institute of Materia Medica has resulted in the identification of a series of compounds related to the structure

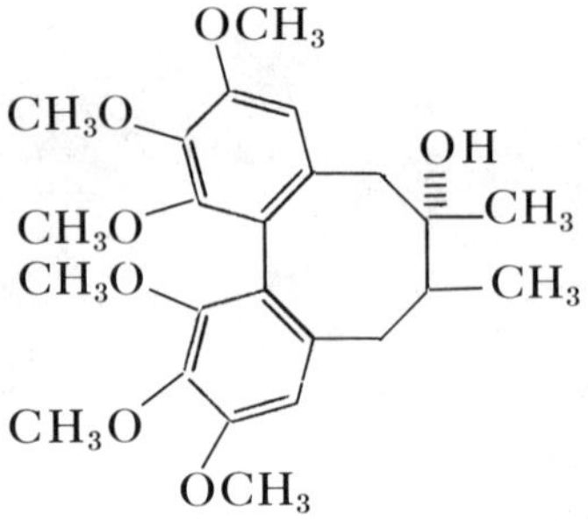

This work was reported in *Scientia Sinica*, **1976** *19*, 276. This compound is in active clinical use in China and is apparently the best known drug for some forms of chronic viral hepatitis.

Chuanghsinmycin, a new antibiotic that was isolated from a microorganism found in a Chinese soil sample, is active against gram positive and some gram negative organisms and is useful in treating urinary infections. The structure

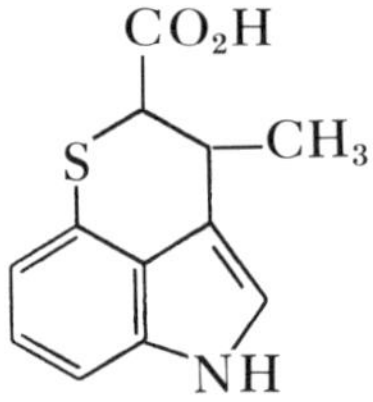

has been determined on the basis of chemical and spectroscopic evidence that was reported in *Acta Chemica Sinica*, **1976** *34*, 129. Finally, a compound

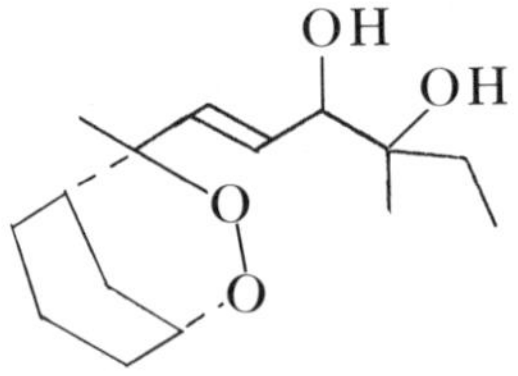

was isolated from *Artabotrys uncinatus*, a traditional medicinal herb, and was determined to have the structure shown by a combination of chemical and spectroscopic studies. It is apparently useful in combating protozoa.

Although the spectroscopic instrumentation in this institute is somewhat routine in character with only a rather ancient 60-MHz NMR machine, its location in Peking makes it possible for the Institute to make use of the 250-MHz NMR machine* at the Institute of

* This instrument is apparently not operational because of difficulties in providing adequate servicing.

Chemistry in Peking and the XL-100 for ^{13}C NMR in the Institute of Photography. Some use is made of these instruments in neighboring institutions, but not nearly so much as would be desirable if good instrumentation existed in the Institute of Materia Medica itself.

Institute of Materia Medica, Shanghai

Among the projects being pursued at this Institute is extensive work on camptothecin

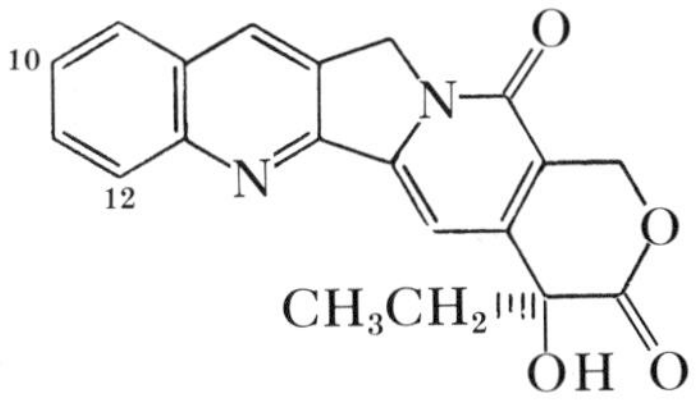

and various derivatives. People at this institute have found that 10-hydroxy-camptothecin as well as 12-chlorocamptothecin have much improved pharmacological properties over the parent compound, which is an anticancer substance of very high toxicity. Because of the increased interest in some of these derivatives, they have also worked out a very clean 10-step synthesis of camptothecin itself which was reported in *Scientia Sinica,* **1978** *21,* 87. This proceeds in an overall 18% isolated yield and seems to be broadly applicable to a variety of derivatives. Work is underway on derivatives of lycorine and of harringtonine.

Work in phytochemistry can be illustrated by the recent isolation of rorifone

$$CH_3SO_2(CH_2)_9C{\equiv}N$$

from a traditional Chinese medicinal herb. This material is useful in the treatment of bronchitis, and it is now used in the synthetic form to avoid the side effects from other components of the herb. Rorifone is also the subject of the normal kinds of studies of structure-activity relationships in this series. A new antihelminthic, quisqualic acid

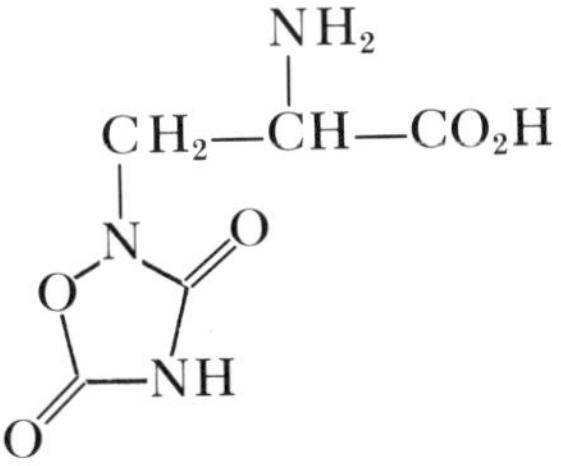

has also been isolated from traditional medicinal plants.

One of the unique opportunities available to Chinese chemists is the wealth of medicinally active substances to be found in their traditional medicines. There are approximately 5000 widely used Chinese medicines, and only about 10% of these have been examined in any serious way by chemists. Although Chairman Mao once said, "Chinese medicine and pharmacology are a great treasure house and efforts should be made to explore them and raise them to a higher level," this exploration was not listed as one of the eight priorities in the recently issued list. However, scientists in the field believe that an emphasis will continue to be made in this area and that the exploration of these traditional Chinese medicines continues to be one of the most exciting aspects of Chinese chemistry.

Institute of Chemistry, Peking

A number of the projects that are going on here deal with areas of applied organic chemistry, and, in particular, polymer chemistry. This includes work on making modified polypropylene and on making various polymides. Some polyquinoxyalines and polytriazines are also under study.

Work is in progress on graph theory of molecular orbitals in an effort to account for the energy levels of simple organic molecules. The relationship between the eigenvalue matrix and molecular symmetry has been explored. It was contended that this approach is easier than using group theory character tables.

There was discussion of studies of organic semiconductors involving complexes of TCNQ (tetracyanoquinodimethane) with TTF (tetrathiofulvaline).

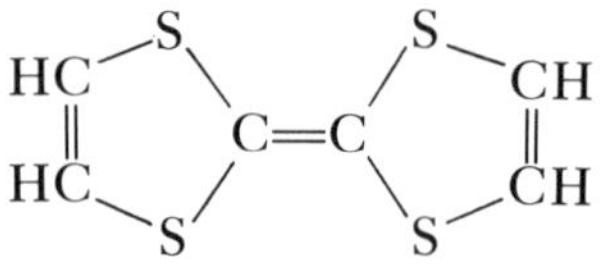

A new approach for synthesizing TTF from acetylene has been devised. TCNQ was obtained using a procedure devised by du Pont. The complex was made by mixing solutions in acetonitrile. Single crystals of the complex were isolated. Two kinds of crystals were found, one of which is highly conductive. The dc conductivity of single crystals (5 mm $\times$ 0.2 mm) was measured as a function of temperature and exhibited a maximum at about 30°C. The conductivity was ohmic up to 500 μa (40 amps/cm^2) in the direction of maximum conductivity. This work has been extended to polymeric semiconductors.

Institute of Biochemistry, Shanghai

This institute has made a significant effort in protein and peptide chemistry in which new synthetic methods are being developed. Members of the institute played an active role in the Chinese insulin synthesis that was completed a number of years ago, and they have continued to study polypeptide synthesis and the properties of various polypeptide hormones including modified insulins. The synthetic strategy involved uses conventional methods to prepare peptides with seven or eight amino acids; these peptides are then carefully purified. Next polypeptides are assembled using an insoluble supporting resin. The result is that many fewer reactions are actually performed on the polymer, and therefore the probability of error and resulting mixtures is smaller. Using this strategy, for instance, glucagon has been synthesized. The 29 amino acids of this polypeptide were assembled by building a tripeptide on a solid support and then adding a pentapeptide, then a hexapeptide, then a nonapeptide, and then a hexapeptide. The overall yield was 83%. Studies with insulin have focused on the possibility that various fragments can be removed or modified with retention of biological activity. Reports on this have been published in *Scientia Sinica.* Syntheses of oxytocin and vasopressin that were developed in this Institute are now in commercial production.

Kirin Institute of Applied Chemistry

A program of interest at this Institute is the study of chemiluminescence processes for the development of practical light sources. The basic compound being used is an organic diester that reacts with hydrogen peroxide to produce the compound $\mathrm{RO\overset{\overset{O}{\|}}{C}-\overset{\overset{O}{\|}}{C}-OOH}$

$$\mathrm{RO\overset{\overset{O}{\|}}{C}-\overset{\overset{O}{\|}}{C}-OR + H_2O_2 \rightarrow RO-\overset{\overset{O}{\|}}{C}-\overset{\overset{O}{\|}}{C}-OOH + ROH}$$

The peroxide then decomposes into an oxalate diester, $\begin{array}{c}\mathrm{O\quad O}\\ \|\quad\ \|\\ \mathrm{C-C}\\ |\quad\ |\\ \mathrm{O-O}\end{array}$ plus

ROH. Next the oxalate diester intermediate reacts with a fluorescent molecule, an organic compound (to form a complex), which then leads to the excited state of the fluorescent molecule plus two molecules of CO_2. The excited state of the fluorescent molecule returns to the ground state with the emission of visible light. The fluorescent molecules are substituted anthracenes.

The fluorescent molecules have large quantum efficiencies per se. The oxalate ester decreases the fluorescence efficiency and its concentration has to be minimized. Experiments of this type have been done previously in West Germany and at Columbia University, and the present program is designed primarily to train personnel in this area. We were shown some of the devices they have produced by using the chemiluminescent reaction. The usable life of one of these sources after mixing the peroxide with the diester and the fluorescent molecule in sealed plastic vials is of the order of three to six hours.

Other Locations

Very limited programs in organic chemistry are also underway in most of the other institutes and universities we visited. A new program in photochemistry is being initiated at the Institute of Photography in Peking, but we did not visit this institute.

General Assessment

Chinese organic chemistry is relatively strong in some areas, particularly those involving medicinal compounds and other natural products of practical interest, such as insect control agents. On the other hand, there has not been much emphasis on the invention of new synthetic reactions, the invention of new methods for structure determination, or the development of the field of physical organic chemistry. Since a new emphasis is to be placed on basic research as an intellectual support for the strong applied research programs that have been pursued, presumably the imbalances will be corrected and broad general programs of research in organic chemistry will be pursued in a variety of institutions.

B. Inorganic Chemistry

Rare Earth Chemistry

If the class of elements denoted "rare earths" had been discovered in China, they would not carry that name. This group of elements is abundant, and almost all chemical research laboratories are looking for applications of rare earths and rare earth compounds. For example, work at Futan University is directed toward the use of rare earth compounds as luminescent materials and also as converters for X-radiation to ultraviolet and visible radiation to increase the sensitivity of X-ray films. Work has been done to find rare earth compounds stimulated either by ultraviolet radiation or low-energy X-radiation and conversion of this radiation to visible light for use in fluorescent lighting. Researchers at Futan University have developed fluorescent coatings of rare earth compounds that achieve a luminescence of 80 lumens per watt in conventional type fluorescent tubes. We were shown three types of coatings: red (which contains Y_2O_3-Eu), green ($Cd_{1-x}Tb_xMgAl_{11}O_{19}$), and blue ($Ba_{1-x}Eu_xMg_2Al_{16}O_{27}$). The fluorescent coatings are used on conventional mercury discharge tubes. White light is produced by mixing fluorescent coatings of various rare earth compounds giving different types of white light equivalent to warm and cool light bulbs in the United States.

A second major application of rare earths is in intensifier screens for the conversion of X-radiation in medical applications. It is now conventional practice in many places, including the United States, to insert a converter in front of the photographic sheet recording the X-radiation after it has passed through a patient. A luminescent screen is used as a sandwich in front of and behind the X-ray film. The X-ray impinges on this sandwich after it has passed through the patient under examination. In this process the X-radiation is converted to visible light by the fluorescent screens both in front of and behind the photographic film, converting the X-radiation to visible radiation, which is much more effective in producing an image on the silver halide film. Conventional practice is to use calcium tungstate ($CaWO_4$). The workers at Futan University have developed fluorescent coatings consisting of La_2O_2S. These coatings have four times the efficiency in converting X-radiation to an effective image on the photographic sheet as does a conventional calcium tungstate coating. In particular, with 65 kV X-radiation and a 100 mÅ current, it usually requires 1.2 seconds of exposure with the calcium tungstate converter. The lanthanum oxysulphide converter produces the same exposure and resolution in 0.3 seconds. The Chinese have taken advantage of this to produce portable X-ray machines that are now being shipped to small villages to make X-radiation facilities available.

In the 1950s the inorganic chemistry group at the Kirin Institute of Applied Chemistry achieved the analysis of rare earth elements by ion exchange. This group is now doing work on separation, extraction, and purification of rare earths, as well as work on the synthesis of rare earth compounds and alloys, particularly semiconductors. They have made compounds of selenium and gadolinium on a one-to-one basis and are producing rare earth doped laser materials. We were shown a laser rod that was garnet doped with yttrium-aluminum. They have the capacity of making the laser rods at the institute and are studying the spectroscopic properties of the laser rods including color centers.

There is also a program in photovoltaic materials at this institute. The principal program is in the area of photocells. One type is a multilayer composite that consists of a CdS-Cu_2S unit. The cadmium sulfide is vacuum evaporated onto a plastic backing. The cuprous sulfide is then deposited in a controlled way by dipping the film into aqueous solution. Next the unit is gold plated. One of these cells can produce up to 450 mV. We were shown units that were 50 cm^2 in area. The present prototype cell is clearly expensive and does not seem to be a practical unit. They are presently studying the stability of the cell, as well as studying polycrystalline silicon photocells. In the future they hope to investigate the possibility of rare earth photovoltaic cells.

There is also a program to investigate gas sensitive semiconductors at the Kirin Institute of Applied Chemistry. These are small devices which are useful for analyzing combustible gases at the ppm level. The selectivity is not too good at present, but the units are already useful for safety detection in mines.

Rare earth separation is done by solvent extraction using complexing agents. Carboxylic acids are the cheapest, but phosphorus complexing reagents have the greatest sensitivity. Amines are also useful for the separation of rare earths as a group from other elements. The light rare earths are separated by solvent extraction. Heavy rare earth separation is by ion exchange using cation exchange resins of the polystyrene sulfonate type. Ammonium acetate is currently being used as the eluting agent, while EDTA was formerly used. Ammonium acetate is a simpler and cheaper reagent for the separation of yttrium. Alpha-hydroxyisobutyric acid is not used in separation operations since it is too expensive, but it is used in analysis. The solvent used in the solvent extraction work is kerosene, since there are strict limitations against the use of aromatics because of health considerations. (This is one of the few instances in which we heard of any health considerations in the environment for laboratory and plant workers.) The solvent extraction is done in mixer settlers.

Organometallic Chemistry

Research in organometallic chemistry in China is still quite limited. Research described at the Institute of Organic Chemistry in Shanghai involved the production of molecular nitrogen compounds of the type $Mo(N_2)_2L_2$ where L is a diphosphine. When some of these complexes are acidified with sulfuric acid, ammonia is generated. This work is directly related to that by Chatt and co-workers. The work has involved thus far L = $Ph_2PCH_2CH_2PPh_2$, which does not yield NH_3, and the unsymmetrical $Ph_2PCH_2CH_2PEt_2$, which does show production of ammonia on acidification. In order to decide which isomer they have made in the unsymmetrical diphosphine, a crystal structure determination is underway.

Organometallic chemistry is often more difficult than organic synthetic chemistry, requiring in many cases the handling of extremely air-sensitive compounds either in Schlenk ware or in dry boxes. It is thus likely that the Chinese effort in organometallic chemistry will lag until their current preliminary efforts succeed in coming up with an important new product or homogeneous catalytic process.

From conversations it appears that metals in short supply in China include cobalt, nickel, chromium, and the platinum group. On the other hand, the rare earths are common.

Chemical Crystallography

Prior to 1950 X-ray crystallography was a useful tool for understanding a variety of minerals and simple inorganic substances, but at that time its application to more complex structures was very limited. It is only in the past 10–15 years with the advent of high-speed computation facilities and automatic diffractometers that chemical crystallography has become a routine method for chemical analysis in the West. The capability for modern X-ray crystallography in China is limited.

The most advanced work in crystallography is being carried out at the Institute of Biophysics in Peking. A form of insulin was synthesized in Shanghai in the 1960s and crystallization and research on the structure of insulin was undertaken at the Institute of Biophysics in Peking. By 1971 they had reached 2.5Å resolution and by 1973 1.8 Å resolution. The Chinese have demonstrated a solid understanding of the solution of protein structures including data collection and the preparation of heavy atom derivatives. However, it would be difficult for them to implement the kinds of model refinement calculations that have been pioneered in the West. The equipment in use is a relatively recent, perhaps 1972 vintage, Phillips diffractometer. This is the only four-circle, automatic diffractometer in China. As such it is kept very busy. It is of primary use to the Institute of Biophysics but it can be scheduled by other institutes.

Although structural chemistry is reportedly one of the main tasks at the Institute of Chemistry in Peking, the crystallographer Wang Shou-tao has available only film cameras; however, he has access to the diffractometer at the Institute of Biophysics in Peking. Wang was working on a 23-atom alkaloid with four molecules in the common noncentrosymmetric space group $P2_12_12_1$. Ninety percent of the time such a structure would be solved in a largely routine fashion in the United States using one of a number of available direct methods programs. In Peking one year was spent to find a trial structure and no refinement has as yet been initiated. Wang apparently chose to write his own direct methods program.

The X-ray diffraction setup at the Kirin Institute of Applied Chemistry, a rather sophisticated institute, is primitive. Two tube stands are available and, in addition to powder units, two Nonius Weissenberg cameras. The group was aware of the four-circle diffractometer at Peking but has not used it as yet. The choice of problems appeared to be up to the individual investigator within the confines of problems of general interest at the Institute. Among the structures done in the past were ethylenediaminetetraacetic acid, a calcium acetylacetonate, a mineral of composition $PbNb_2O_6$, and a praeseodymium tris-acetylacetonate. The structure discussed with us was that of 3,3,3-nitrilo-trisproprionamide. This apparently is a so-called bond transfer agent used in polyacrylamide synthesis and in waste water treatment probably because it is a good complexing agent for certain metals. As described, the material is monoclinic, space group C_{2h}^6–C 2/c. Apparently data had been collected by oscillation methods around the three axes, *a*, *b*, *c*, and intensity statistics had been run that indicated the material to be in the centrosymmetric space group C_{2h}^6. In answer to a question whether with four molecules in the cell symmetry would be imposed on the individual molecule and whether such symmetry was impossible without disordering, there followed an inconclusive discussion. Apparently it had taken approximately a year to solve each structure but the trisproprionamide had not yet been solved.

The X-ray equipment at the Institute of Organic Chemistry in Shanghai involves a powder diffractometer, a unit for low-angle X-ray scattering, and a Nonius Weissenberg camera. In the particular case of the Mo dinitrogen complex mentioned under "Organometallic Chemistry," the unit cell data had been collected in Shanghai, but the diffraction data had been collected at the Institute of Biophysics in Peking. The space group of this particular material could either be P2/c, which is a centrosymmetric group, or Pc, which is noncentrosymmetric. We were told that it was definitely in Pc as a second harmonic generator (SHG) test was positive. SHG is a very new technique involving laser technology, and it is interesting to learn that it had been adopted in China. This work is being done at the Institute of Ceramic Chemistry and Technology where its use is to screen compounds that might have piezo-electric and other properties that depend upon the absence of a center of symmetry.

If the Chinese are to develop viable programs in modern inorganic chemistry, the routine use of crystallographic structure determination will be essential.

C. Physical Chemistry and Chemical Physics

The principal areas of activity in physical chemistry in China are catalysis and polymer science. These fields dominate the programs in physical chemistry at almost every institute which we visited.

The effort in heterogeneous catalysis is highly applied. It is concerned essentially with (1) process variable studies on existing catalytic systems, (2) the improvement of catalysts for existing processes, and (3) the development of catalysts for processes that the Chinese wish to commercialize. Standard instrumentation for heterogeneous catalyst surface characterization was evident at many of the institutes visited: BET mercury porosimeters, X-ray diffraction, and scanning electron microscopy. However, advanced tools such as ESCA (electron spectroscopy for chemical analysis),* Mössbauer, LEED (low-energy electron diffraction), Auger, or EXAFS (extended X-ray absorption fine structure) are not in use. Also surprising was the lack of emphasis on detailed reaction kinetic studies involving model compounds, and the limited effort devoted to mathematical modeling of catalytic reactions.

* An imported ESCA spectrometer is reported to be available in the Institute of Organic Chemistry in Shanghai, but its use was not discussed.

Polymer efforts in China are comprehensive but not particularly novel. Molecular characterization of polymers is through the usual spectroscopic techniques of infrared and NMR, primarily using imported instrumentation. Solid-state studies involving X-ray diffraction are undertaken usually using imported equipment. Again, there seems to be little activity in theory. No work on chain configuration or thermodynamics was encountered.

In the areas usually grouped under chemical physics in the United States, the most vigorous activity appeared to be in laser chemistry. Many institutions have either just started or plan to start some programs in laser chemistry. At present most of the laboratories are in the stage of laser construction and have not reached the stage of involvement in active research on laser chemistry problems. In many of the laboratories visited, it was evident that the ability of Chinese experimentalists to get lasers running is excellent.

At present there are no molecular beam research programs in China, although the Institute of Chemistry in Peking, the Institute of Chemical Physics in Talien, and perhaps the University of Science and Technology in Hofei are planning to initiate programs in microscopic chemical kinetics in the near future.

Active research programs in molecular spectroscopy for the determination of molecular structure involving, for example, microwave, infrared, visible, and UV spectroscopy in the gas phase are largely absent. There are programs of high quality at a number of institutions in molecular orbital calculations, but theoretical chemistry, particularly activities involving large-scale molecular calculations, does not appear to be part of the basic research activity in chemistry in China. We also did not see any active research in the field of statistical mechanics; however, in China this field is apparently considered to be a subdiscipline of physics instead of an interdisciplinary field in chemistry and physics.

D. Nuclear Chemistry

Research in nuclear chemistry covering the areas of low-energy nuclear physics and radiochemistry is conducted at the Institute of Atomic Energy, Peking; the Institute of Nuclear Research, Shanghai; and the Institute of Modern Physics, Lanchow. The first work in this field started at a nuclear physics institute, known as the Institute of Modern Physics, that was formed in Peking in 1950. This institute served as the source of the three institutes devoted to nuclear science. The parent institute evolved into the Institute of Atomic Energy and furnished personnel for the start of the Institute of Nuclear Research and the Institute of Modern Physics. Each of the three institutes has a cyclotron, other low-energy accelerators, and radiochemistry laboratories, and the Institute of Atomic Energy has two research reactors. Theoretical work in the field of high-energy nuclear physics has been associated with the Institute of Atomic Energy, but recently experimental work in the high-energy nuclear physics field has been centered in a newly created Institute of High-Energy Physics in Peking where there are plans to build a high-energy proton accelerator.

Institute of Atomic Energy, Peking

The general programs of the Institute of Atomic Energy are the following: reactor material testing, fuel reprocessing, theoretical nuclear physics, neutron physics, work with two accelerators, an electronics laboratory, theoretical heavy-ion physics, measurement of neutron cross-sections, and inelastic neutron scattering. The major facilities available at the institute are a 1.2-meter diameter cyclotron for the acceleration of protons and deuterons, a 600-keV Cockroft–Walton D-T accelerator to produce 14 MeV neutrons, a 7-mW heavy water reactor containing 2% uranium-235 built by the U.S.S.R., and a light-water swimming-pool reactor containing 3% uranium-235. Construction of the swimming-pool reactor started in 1960, and it was put into operation in 1964. They do not claim that any of their experimental facilities are at the forefront of modern nuclear science.

The chemistry program includes radiochemistry, analytical chemistry, waste water treatment, fuel reprocessing, plutonium chemistry, and analytical chemistry. Under analytical chemistry, they have programs in neutron activation analysis, charged-particle and fast-reaction activation analysis, and proton-induced X-ray fluorescence. Their work on activation analysis includes the study of rare earths in meteorites. There is also a program in transplutonium chemistry. They have programs in high-pressure ion exchange, production of radio pharmaceuticals, and do a limited amount of tracer studies for medicinal chemistry.

The transplutonium program consists of a study of americium–curium separation. Americium-241 is produced somewhere in China. Samples of this americium are irradiated at the Institute for

Atomic Energy in the neutron reactor to produce curium-242 at a multimicrocurie level. The separation method under study involves cation ion exchange resins. The resin is a sulfonic acid exchange resin made in China. The resin size is about 15 micron and the resin is loaded in a column 45 cm long by 2 mm in diameter. The column is stainless steel and the elution velocity is 0.075 cc/ min. The complexing agent is α-hydroxyisobutyric acid. In order to reduce the interference from gas bubbles produced by the radiolytic decomposition of the resin, they have gone over to high pressure ion exchange at approximately 20 atmospheres (20 kg per cm^2). Their equipment can go to 200 atmospheres. The separation is studied by an on-line sodium iodide gamma detector shown in Fig. 9. They do have some Ge-Li detectors, but these are probably in short supply. The separation spectra we saw showed a resolution of americium and curium, but just resolution. The peaks were rather broad. The gamma spectra are analyzed on an 800-channel multichannel analyzer, which is manufactured by Intertechnique in France. All the other equipment used in the gamma ray counting is made in Peking. Data are finally recorded on a paper tape printer made in France. The sodium iodide detector is operated on-line. Alpha spectra are also measured on-line by gold-

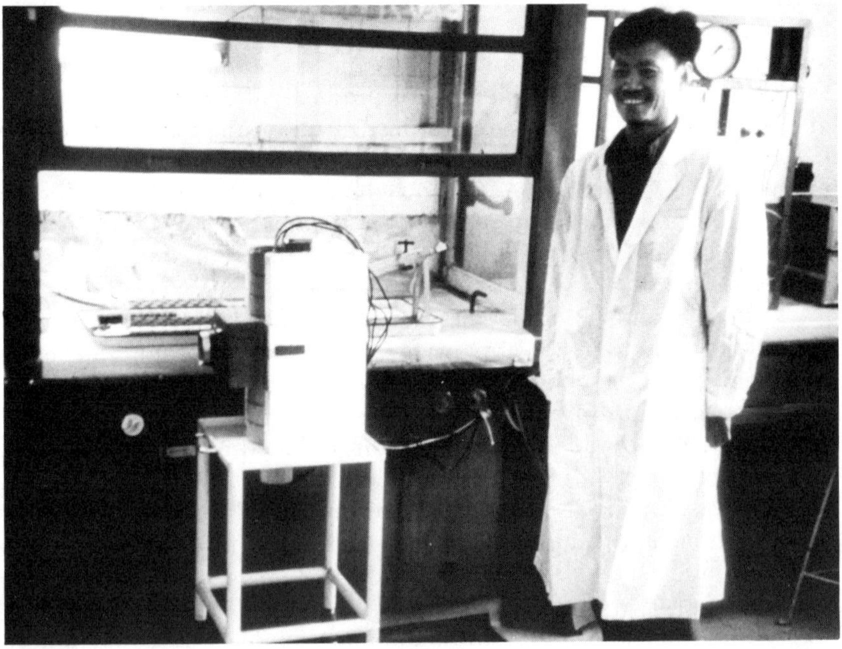

Figure 9. Actinide Chemistry Laboratory, Institute of Atomic Energy, Peking. Ho Chang-Yu, Deputy Director, Radiochemistry Laboratory.

silicon solid state detectors. A very thin film, approximately 8 microns thick, flows past the detector. The alpha peaks are barely discernible above background.

Radioisotope production is under the direction of Dr. Hsiao Yi-chung, who worked with R. B. Duffield at the University of Illinois from 1947 through 1951. This facility is the major producer of radioisotopes in China. The principal production facilities are for the production of ^{198}Au, ^{51}Cr, ^{131}I, ^{125}I, ^{131}Cs, and ^{32}P (carrier free). The principal production is ^{131}I and ^{198}Au. Each of these isotopes is processed in a batch process of about 40 curies. The facility was constructed in 1970 and put in full operation in 1972. Some of the equipment was built in the U.S.S.R.; the recent equipment was built in China.

They also have facilities for the preparation of ^{14}C and T labeled compounds. ^{14}C is prepared by reactor irradiation of $Ba(NO_3)_2$ or AlN. The final product is barium carbonate $Ba^{14}CO_3$. The specific activity of the ^{14}C is 20 millicuries per millimole $BaCO_3$. ^{14}C labeled sodium benzoate and sodium cyanide are also prepared. They do not perform any complicated organic synthesis with ^{14}C at this facility.

They have an interesting and significant program of tritium labeling of the active ingredients of Chinese herbs. Some of the compounds being labeled are scopolamine, avacoline, and harringtonine. Tritium labeling is done by a modification of the Wilzbach method. In addition to exposure to T_2, the T_2 gas is discharged to induce an exchange. They also use catalytic addition to the double bond or isotope exchange in their tritium labeling.

This institute is interested in obtaining an electrostatic accelerator, either an Emperor Tandem from High Voltage Engineering Corporation or a Pellatron from NEC.

Institute of Nuclear Research, Shanghai

Nuclear physics research at the Institute of Nuclear Research deals mostly with nuclear reactions and the study of fission. Isotope research is primarily concerned with radioisotopes produced in the cyclotron and tritium-labeling. The institute has a ^{60}Co source of 120,000 curies for radiation research purchased from AECL (Canada), as well as a 1.2-meter diameter cyclotron (shown in Fig. 10) that was built in the Institute. There are plans to build a Tandem accelerator for 12 MeV protons. This accelerator is intended to be similar to the FN accelerator that is marketed by High Voltage Engineering Corporation. The accelerator is currently in the design stage, and a scientific program for the use of the accelerator is in the process of being set up. It is

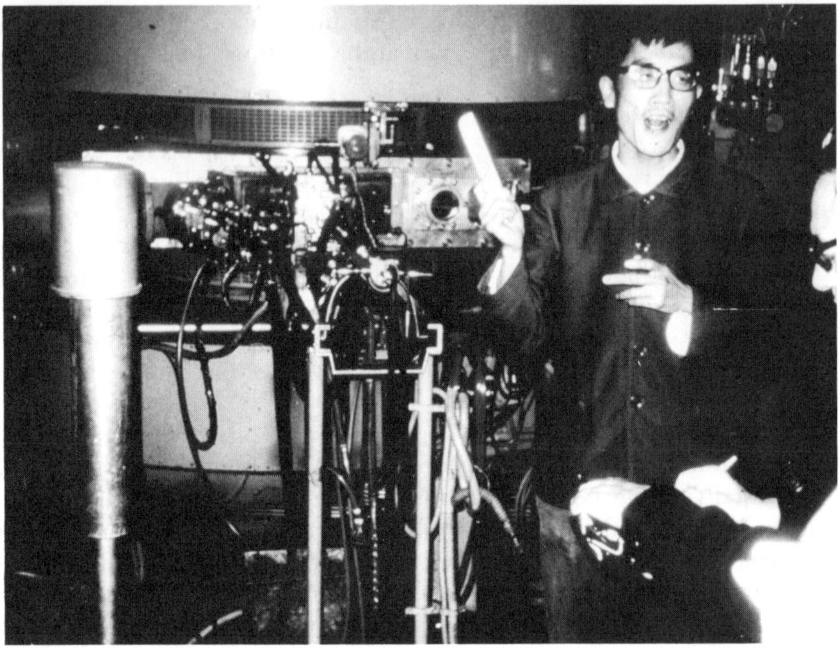

Figure 10. 1.2-meter cyclotron, Institute of Nuclear Research, Shanghai.

anticipated that the work with the Tandem accelerator will be application-oriented, and it is not intended to be a forefront instrument.

The cyclotron is conceptually similar in design to the cyclotron at the Institute of Atomic Energy in Peking built by the Soviets. The cyclotron here, however, was completely designed and built by the Shanghai group. The design started in 1960 and actual operation commenced in 1964. The accelerator was designed for 25 MeV α-particles, and it actually produced 30 MeV particles. It also accelerates deuterons and protons. The proton energy has now been increased to 8 MeV. The machine is in the process of being converted to a variable energy sector-focused machine to produce 10–30 MeV protons, 20–60 MeV deuterons, and also heavy ions. They hope to have a 1 mA beam at the target. The present internal beam has a maximum intensity of 1 mA for protons. They bring out an external beam of 80 μA of deuterons and get good three-dimensional steering by the use of a combination of quadrupole and bending magnets. Presently the cyclotron is being used to produce radioisotopes for nuclear medicine: ^{67}Ga, ^{85}Sr, and ^{111}In. They are also using the proton beam to do X-ray induced fluorescence. There is also a cooperative program between scientists at the Institute of Nuclear Physics and physicists at Futan University who are doing proton induced X-ray fluorescence.

Of special interest was a program to analyze valuable old swords dating as far back as the Yu State (about 2500 years ago) by this method. They have found that the body of the sword consists of the elements copper, tin, iron in small amounts, and lead. They have analyzed green lines on the surface of the sword and find that these contain sulfur. The handle of the sword is inlaid with glass containing, according to their X-ray fluorescence analysis, silicon, potassium, and calcium. In this work, the proton source is the electrostatic generator at Futan University.

The major work here in radiochemistry is the preparation of tritium-labeled compounds and the study of their radiation stability in storage. Radiolytic decomposition is followed by paper chromatography and electrophoretic analysis of the tritium-labeled material. They are also studying the effects of beta radiation and mass effects of tritium on biological compounds, particularly the cells of the thyroid. All of this work is being done in lucite glove boxes. They have a program for tritium-labeling of nucleosides: ADT, UDP, and cytosine including the synthesis of material at a level of 20 curies of tritium per millimole of nucleoside. The labeling is being done by chemical and biological synthesis starting with the tritium-labeled amino acids. The tritium label is introduced by reduction of unsaturated compounds using tritium gas and various catalysts. They are also tritium-labeling steroids and hormones including cortisone, progesterone, and testosterone. They have a very fine facility for handling tritium and ^{14}C similar to the Argonne and Los Alamos type hoods. At the time of our visit to the ^{60}Co source, there was little scientific work in evidence.

The Institute of Nuclear Research in Shanghai has a facility for producing semiconductor detectors, including Ge–Li coaxial-type detectors for gamma rays and Si–Au surface barrier detectors for charged particles. The germanium is obtained from the Institute of Nonferrous Metals in Peking, a very large institute that we did not know about beforehand. In this department they are producing plug board integrated circuits as shown in Fig. 11 and have already built 512-channel analyzers. These were the first Chinese-built multichannel analyzers with integrated circuits that we saw on our visit to China.

The institute has a 14-MeV neutron generator. The neutrons are produced by the D-T reaction.

There were three types of perturbed angular correlation experiments going on. One was the measurement of the magnetic moment of the first excited state of cadmium-111, which is produced from indium-111. A second area of work was the use of the cadmium-111 isotope to measure local magnetic fields in solid-state materials.

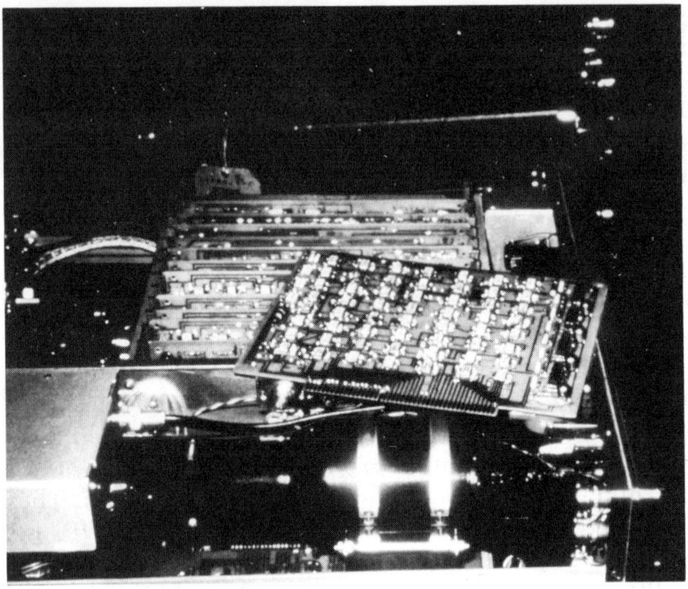

Figure 11. Plug board integrated circuits used in Chinese-built 512 channel multichannel analyzer.

The cadmium-111 was apparently implanted by recoil. Perhaps the most interesting experimental project was the application of perturbed angular correlation to study the enzyme nitrogenase, which is responsible for nitrogen fixation. The indium-111 isotope is bound to ATP, which is directly associated with the enzyme. It is not apparent that this rather nonspecific binding of indium can provide much useful information on the rotational correlation time for the enzyme. This group is evidently working in collaboration with the Institute of Plant Physiology in Shanghai. They had also proposed to use the isotope ^{99}Mo in these studies since molybdenum is associated with iron in the active site of the enzyme. The principal investigator responsible for this work was Ni Hsing-po. Dr. Ni had studied nitrogenase that had been exposed to atmospheres of different gases at 13°C for periods of two hours and indicated that he saw differences in the perturbed angular correlation spectra for different mixtures of gases. Although the angular correlation experiment appeared to be competently done, the experimental strategy did not appear likely to provide information of biological interest.

When Professor Seaborg visited the Institute in 1973, work on nuclear thermoelectric power sources was in progress. The aim was to develop a small power source that used the heat from the radioactive decay of α particle emitters. This heat would be converted to electric power through the thermoelectric process. This work has been discontinued at the Institute of Nuclear Physics. Such power sources are still in a design stage, and work is going on at another site.

Institute of Modern Physics, Lanchow

The Institute of Modern Physics at Lanchow, since its inception in 1957 and its operational start in 1963, has been under the direction of Yang Ch'eng-chung, who did his graduate work with James Chadwick at Liverpool, England. The main research facility is an accelerator for light heavy ions (C, N, O), a 1.5-meter cyclotron whose construction was started by the Soviet Union in the late 1950s and was completed by the scientists and engineers at the institute when the Soviets abandoned them in 1960. The cyclotron, beginning in 1963, was used to accelerate protons, deuterons (24 MeV), and helium ions, and was converted to the acceleration of heavy ions in 1973.

The department for the "Design of New Accelerators" is working on the design of a recently authorized heavy ion accelerator, a tandem cyclotron system, in which the present 1.5-meter cyclotron, redesigned as a sector-focused machine, will serve as an injector for a 6.15-meter sector-focused cyclotron. When this is completed in the early 1980s, it will have the capability to accelerate light heavy ions (such as carbon) to an energy of 50 MeV per nucleon and heavier ions (such as xenon) up to energies of 6 MeV per nucleon. (A future, second phase construction utilizing a better injector might give the capability of accelerating even heavier ions (e.g., U^{+36}) to an energy of 10 MeV per nucleon).

There are two magnetically directed beam lines entering the experimental area, as well as a zero degree line entering the chemistry area, equipped with a helium jet system to transport recoil product nuclei to γ ray (Ge–Li) and α particle (Si–Au) detector systems. The radiochemists have identified actinide products (^{243}Cf, ^{244}Cf, ^{246}Cf, ^{246}Es, ^{250}Fm) produced in the reactions of ^{12}C with ^{238}U and ^{239}Pu target nuclei, based on chemical separations employing a cation exchange resin and elution with α-hydroxyisobutyrate ion. In the process of eliminating the adverse effect of lead impurity (which results in the production of interfering isotopes of Po, At, Fr, etc.) in such investigations, they have developed an activation-analysis method that is sensitive to the detection of Pb in the atmosphere at concentrations as low as 10^{-8}–10^{-9} g per cubic meter.

Their radioanalytical method for the determination of the yields of products from heavy ion bombardments, in which independent and partial cumulative yields are summed in order to give total isobaric yields, has been applied to the measurement of the symmetric yields of fission products produced in the bombardment of Au, Bi, and U targets with ^{12}C ions. This approach has been supplemented by the use of mica, plastic, and glass detectors to measure the values for fusion–fission cross-sections in these and other reactions. The program includes the measurement of a number of cross-sections for the production of evaporation residue products such as the production of a range of neutron-deficient iodine isotopes produced in the reaction of ^{12}C with Ag target nuclei.

There is also a program of on-line, kinematic measurements, using the E, ΔE energy measurement technique, to identify projectile-like products and their angular and energy distributions resulting from heavy-ion reactions. They have investigated the characteristics of Li, Be, and B isotopes produced from the reaction of ^{12}C with ^{209}Bi target nuclei. In the reaction of ^{12}C with ^{27}Al target nuclei, they have identified products ranging in atomic number from He to Si.

The "applied nuclear physics" department includes the work on radiochemistry. Numerous isotopes useful in medical and other practical applications, produced in the cyclotron and from other sources, are isolated here—the list includes ^{7}Be, ^{22}Na, ^{51}Cr, ^{55}Fe, ^{59}Fe, ^{56}Co, ^{57}Co, ^{68}Ge (source of ^{68}Ga), ^{85}Sr, ^{109}Cd, ^{181}W.

The institute has a Cockcroft–Walton Accelerator shown in Fig. 12 with an operating voltage of 400 kv, which is used to produce 14

Figure 12. Cockroft-Walton accelerator, Institute of Modern Physics, Lanchow.

MeV neutrons at a yield of 3×10^{11} neutron/sec from the D-T reaction. These neutrons have been used to make a careful determination of the mass distribution, with the aid of chemical separations, of the fission products from uranium.

The "electronic, detectors, and computers" department is housed in a separate building. Here they produce Si surface barrier detectors, ion implantation detectors, and Li-drifted Ge detectors for γ rays. The computer here, built in Harbin, is programmed for Fortran II and has only a moderate capability.

E. Analytical Chemistry

Examples of programs in analytical chemistry in progress at several of the institutes visited include:

Talien Institute of Chemical Physics

Work in gas chromatography was started at the Talien Institute of Chemical Physics in 1954. By 1960 they had done a significant amount of work using kieselguhr support—80 to 100 mesh. The work has been done with glass capillary columns and thermal conductivity and flame ionization detectors. These units were put into production and cuts from petroleum fractionation were studied. They established the relationship between the boiling point of various petroleum fractions and the retention times on the glass capillary columns and simultaneously worked on the dynamic theory of chromatography.

Subsequently developed at this institute were microanalytical methods for organic compounds by gas chromatography including analysis for nitrogen, chlorine, sulfur, and phosphorus directly by gas chromatography. Important problems have been elimination of the catalytic activity of the support and elimination of adsorption on the chromatographic bed. In some cases they convert the compound to be analyzed chemically to another form to minimize adsorption. They have made major advances in the development of flame ionization and electron capture detectors including new detectors that can discriminate selectively for nitrogen and phosphorus in organic compounds. These detectors have a selectivity against hydrocarbons of the order of 10^4. The sensitivity of their detectors is of the order of a picogram per second. The response of the detectors has been optimized by the detailed study of the effects of the flow rates and composition of hydrogen, oxygen, and the carrier gas. The carrier gas is nitrogen prepurified by molecular sieve, activated charcoal, and silica gel. They can completely suppress a hydrocarbon peak in this system while analyzing for nitrogen and phosphorus. They can detect 10^{-13} grams of DDT per second. The real limit of sensitivity is a nanogram of material per milliliter of solution. They require 10^{-3} mL of liquid solution to make an analysis. Therefore, they can really detect on the order of a picogram. They expect to mass produce these detectors. The detectors are being used with British Unicam programmed chromatographic equipment.

They have also developed chromatographic methods for the analysis of impurities in rare gases krypton, xenon, and neon. They can detect impurities at thc 100-ppm level in those rare gases and one ppm of oxygen in argon, which is as good as anybody else can do. Their method of analysis of impurities in rare gases consists of concentration of the impurities on molecular sieves at liquid nitrogen temperature. (Their difficulty in going to lower levels may arise from their inability to trap impurities at temperatures between liquid nitrogen and dry ice temperature.) They interface their GCs directly with a mass spectrometer (an Atlas CH-4) without using any kind of carrier stripper because of the very low flow rate. A typical Chinese-made gas chromatograph is shown in Fig. 13.

Work on high performance liquid chromatography is also in progress at this institute. Their units achieve 10^4 theoretical plates per meter.

Figure 13. Chinese-made gas chromatograph at the Institute of Chemistry, Peking.

Futan University

In addition to conventional work on polarography for dissolved cations in electrochemical analysis, work at Futan University has concentrated on anodic stripping voltametry (ASV). In connection with the latter, they have also carried out a significant instrumental research program. The principle of ASV is to use an electrochemical cell with a platinum anode and a glassy graphite cathode (which is produced by pyrolyzing a high polymer at high temperature). Metal ions dissolved in the electrolyte solution are electroplated on the cathode for approximately one minute. Then the applied potential is decreased linearly and the metal ion redissolves in the solution. The potential at which the metal ion redissolves is characteristic of the metal, and the current that flows is proportional to the concentration of the particular ion. By this method, they can determine dissolved Cu^{2+}, Zn^{2+}, Pb^{2+}, Cd^{2+} in waste waters. The general level of detection achieved for these ions is of the order of 0.1 parts per billion by weight. The various reagents used in the analyses are purified by ASV.

In the area of instrumental development, they have devised instruments that use a gold film electrode for measuring arsenic, tellurium, and selenium by ASV. These instruments have sensitivity of less than 0.1 parts per billion. The instruments have been designed by the advanced students in the program (both third-year undergraduate and graduate students) and have been put into mass production by an instrument factory in Shanghai. These rugged field instruments can make measurements down to the range of 10 parts per billion to 0.1 parts per billion. This particular program involves 2 of the 10 graduate students in chemistry at Futan University.

An analytical method called "Dissolved Organic Matter by Total Oxygen Demand" is used to determine dissolved organic material (pollutants) in waste water by measuring the oxygen consumption in the combustion of a sample of the water. A small amount of water is injected into a quartz tube by means of a syringe that penetrates a septum. The quartz tube is heated in an oven and contains a platinum catalyst. Nitrogen gas containing a small amount of oxygen flows over the platinum catalyst and carries with it the vaporized water sample. The organic material is oxidized to CO_2. The actual amount of organic material is measured by a differential determination of the oxygen content of the nitrogen stream entering and leaving the combustion tube. The differential oxygen determination is carried out by measuring the potential difference developed between two zirconium oxide electrodes. The sensitivity of the method is moderate: one can determine consumption of 20 milligrams of oxygen per liter of water.

Other Programs

Other programs of interest include work on crown ethers at the Lanchow Institute of Chemical Physics. This program consists of the synthesis of known crown ethers. They have also made an oligomer, or polycondensate, of dibenzo-18-crown-6-formaldehyde. They have made an electrode using this material and have shown that it has a very high selectivity for sodium and potassium and is in fact a potassium ion selective electrode. Apparently their objective is to try to use these crown ethers and cryptates in the separation of lutecium and other rare earth metals. An induction coupled plasma source for emission spectroscopy has been developed at the Kirin Institute of Applied Chemistry. The spectrograph is a Hilger photographic instrument and is used for the determination of rare earth elements.

F. Chemical Engineering

Although there is a great deal of applied research and development in China in direct support of processes that are already underway, there is little basic research in chemical engineering in support of longer-range goals and objectives. Three major areas of applied research in petroleum and petrochemicals, catalysis, and polymers are discussed in detail in Chapter 6.

Basic research in chemical engineering in the United States consists mostly of applied thermodynamics, catalysis and kinetics, reaction engineering, polymers, surface and colloid chemistry, electrochemistry, transport phenomena, fluid mechanics, separation processes, mathematical modeling, process dynamics and control, optimization, and computational techniques. In China many of the applied chemical disciplines are still carried out in departments of chemistry. This is in contrast to the United States, where most of these topics are moving to departments of chemical engineering. Thus, a great deal of the work that would be described as basic research in chemical engineering in the United States is found in chemistry departments and in specialized research institutes in China.

There is a great deal of interest in catalytic techniques and in polymers. There is very little interest in kinetics; the emphasis is in developing catalysts and testing performance. We saw very little of the other topics of research in chemical engineering. There is reported to be some research in separation by reverse osmosis and in reaction engineering at the Talien Institute of Chemical Physics. There is also a Chemical Engineering and Metallurgy Research Institute in Peking, which was discovered too late to be visited. We did meet the Director, Kwauk Moo-son, who received his Ph.D. with Wilhelm at Princeton. Many institutions are working on catalytic cracking catalysts, on ethylene oxidation to ethylene oxide, and on polymerization by Natta-Ziegler catalysts. It is not clear whether or not they are familiar with each other's work and have arrived at a division of labor to avoid unnecessary duplication.

There has been very little published in the research literature during the past 17 years; there are also very few scientific conferences where research results from various research institutes and universities are compared. It is said that the top journal for publication of chemical engineering articles is *Scientia Sinica*, which is a general journal of high-level quality. Since this journal accepts articles from all areas of science and engineering, the number of chemical engineering articles that appear must be very small. There was a journal called Hua-hsueh-kung-bao (*Chemical Engineering Journal*), but its publication was stopped during the time of the Gang of Four. This journal may be revived. Other important research journals include Hua-hsueh-bao (*Chemistry Journal*). There are also two review journals: Hua-hsueh-tong-bao (*Chemistry Communications*) and Ke-hsueh-tong-bao (*Science Communications*).

6. STATUS OF RESEARCH IN KEY AREAS OF TECHNOLOGY

A. Petroleum and Petrochemical Research

The Chinese have an extensive history of petroleum refining, and since Liberation they have developed an increasingly impressive technology based on their earlier experience. Evident are carefully coordinated research efforts, as well as plants and refineries that are almost entirely of Chinese design and construction. On the other hand, the petrochemical area does not enjoy a long history in China, and here the Chinese have imported technology in an effort to catch up as rapidly as possible.

Petroleum and petrochemical research is concentrated at institutes of petroleum and petrochemistry located at either refineries or petrochemical complexes or in the local provinces. There is an expressed desire to locate the research and development functions as close as possible to plants in order to maximize the efficiency with which local problems are identified and solved. In addition to location close to plants, institutes are also located in each province providing support for other provincial industries. The overwhelming fraction of the research effort is applied. Problems are apparently developed on four levels: problems that fit into the long-term needs of the state, problems that are initiated at the plant, problems that are generated by the institute staff, and local provincial problems. Problems that are generated by the institute staff are apt to be highly applied rather than long-range or basic. At no institute did we observe the equivalent of a large long-range research division.

Table I gives an indication of the regionalization of the petroleum and petrochemical research function based on the institutes visited.

There is evidence of duplication of research programs. For example, there is research on zeolitic materials underway at most of the research institutes visited and butene dehydrogenation process studies also appear to be a favorite study area. However, it is difficult

TABLE I
Geographic Spread of Petroleum Institutes
Visited by the U.S. Delegation

1.	Petrochemical Research Institute	Peking
2.	Institute of Chemical Physics	Talien
3.	Academy of Science, Technology and Design	Tach'ing
4.	Refinery Institute Research Laboratory	Tach'ing
5.	Futan University Catalysis Group	Shanghai
6.	Academy of Lanchow Chemical Industry	Lanchow
7.	Institute of Petroleum Research	Lanchow
8.	Institute of Petrochemistry	Harbin
9.	Institute of Organic Chemistry (Petroprotein)	Shanghai

to say whether such duplication is healthy and competitive or redundant and whether the Chinese could be more effective in consolidating their petroleum/ petrochemistry research programs. While duplication exists, differentiation also exists and perhaps dominates in the sense that each plant and province has its own unique problems. The research institute at Harbin, for example, does research on adhesives. Harbin is close to Tach'ing and adhesives are made both at Tach'ing and in experimental quantities at Harbin. The institutes appear to be working on adhesives, in part due to the local production of strain gauges, which require adhesives to cement lead wires to the strain gauges.

There are many examples of the close working relationships between the institutes and the plants. Some examples include:

The use of adhesives at the Institute of Petrochemistry, Harbin, to cement alloys to the end of the drill bits used in the Tach'ing Oil Fields (Fig. 14).

Research at the Peking Petrochemical Research Institute aimed at increasing the activity and stereoregularity of the polypropylene catalysts used in the Peking Petrochemical Complex.

Research at Tach'ing Refinery Institute Research Laboratory aimed at improving the photo-oxidative stability of lubricating oil additives.

Butene oxydehydrogenation research aimed at better process control and improved catalysts for the production of butadiene at the Lanchow Institute of Petroleum Research.

In general the institutes are reasonably well-equipped for their applied function, and process research equipment spans the gamut from small scale microreactors to rather large-scale pilot plants, for example, a 10-liter/hour riser catalytic cracker at the Peking Petrochemical Research Institute and a 1300-liter polypropylene polymerization autoclave at the Futan University catalysis group. The most obvious gap is the lack of computerization and microprocessing so evident in U.S. research laboratories. There is a heavy reliance on GC (including capillary column and stationary phase) and the GC units are invariably produced in China. This internal production of GC units is probably a result of the philosophy of self-reliance with a concomitant difficulty in assembling infrared spectrometers, mass spectrometers, and other specialized electronic and optical instruments. On the other hand, there has been some purchase of foreign equipment in the past 10 years, and the institutes are generally well-equipped with infrared spectrometers, mass spectrometers, X-ray diffractometers, DTA and DSC, X-ray fluorescence, and atomic absorption spectrometers.

Communication between plant and institute seems good, and all institutes and plants visited claimed that people were transferred both from plant to institute and vice versa. Individual projects at times utilized a team approach by institute and plant personnel, and the term "task force" was used several times to describe the approach to problem-solving. Communication exists between institutes on a problem-by-problem basis. Most of this communication is by phone and publication of results is not widespread. There are irregularly sched-

uled meetings between institutes on topical subjects, but the general impression was that the communications network in this field could be improved.

Examples of research programs at specific institutes are given below.

Petrochemical Research Institute, Peking

Research at the Petrochemical Research Institute in Peking is divided into the following categories:

Crude evaluation and processing, including crudes from Tach'ing, Shengli, Takung, Karmanli, and Lungli

Process research and development, including catalytic cracking, reforming, lubricating oil processing and lubricants, solvent deasphalting and dewaxing, solvent refining, and lubricating oil hydrogenation

Catalyst research, with emphasis on research and development on processing catalysts

Products research, including applied research on lubricating oils, greases, and additives

Product evaluation and specification, including engine tests

Analytical research

Instrument development and automation

Apparently this institute had a leading role in introducing fluid catalytic cracking to China. It was stated that in 1959 there was a Russian-designed TCC unit in Lanchow but that the Peking Petrochemical Research Institute designed and had constructed the lower-cost, more effective fluid-bed catalytic cracking unit in the Fushun Refinery. This group also worked out manufacturing procedures for amorphous silica/alumina catalysts and in recent years has developed sieve-type catalysts and a riser cracking unit. The sieves and the riser were commercialized in 1975. They developed a platinum/alumina fixed-bed reforming process based on the initial catalyst research carried out at the Institute of Chemical Physics in Talien. The scaling work and process commercialization was done at Tach'ing in 1965. They began research on bimetallic reforming systems in 1969–70 and now claim to be producing semicommercial quantities of bimetallic reforming catalyst. They appear to be concentrating on a $Pt/Sn/Al_2O_3$ system. It was stated that they, in collaboration with other petroleum institutes, have developed commercial hydrogenation catalysts for coker gas oil, coker gasoline, and lubricating oil hydrogenation (cobalt and nickel-molybdenum). In 1965 the institute developed a two-stage propane deasphalting process with supercritical solvent recovery. This process was commercialized in 1967.

Institute of Petrochemistry, Harbin

Some of the specific projects observed at the Institute of Petrochemistry in Harbin include:

Air oxidation of *o*-xylene to phthalic anhydride over a V_2O_5 carborundum catalyst. This work began in 1970 and a pilot plant was built in 1972, which resulted in a commercial process in 1974. The process uses a $NaNO_3$ salt bath as a heat transfer medium with 3000 reactor tubes and has a molar efficiency of 83%.

Process variable work on the oxidation of 1,2,3-tetramethyl-4-isopropyl benzene to pyromellitic dianhydride over the same V_2O_5/silicon carbide catalyst. The starting material is pseudocumene from coal tar, which is then alkylated prior to oxidation. The Chinese now want to make the pyromellitic dianhydride from Tach'ing bottoms. The ultimate use of the material will be as a monomer for polyimide polymers.

Research on polyacetaldehyde, which is produced from acetamide using a protonic acid with an emulsifying agent (CH_3COBr) as a catalyst. The Chinese use polyacetaldehyde as a replacement for AgI in cloud seeding. They claim it is 1000 times better than AgI and it is, of course, cheaper. This work, sponsored by the Agricultural Ministry, is to develop a cheap, commercial synthesis.

Adhesive research. This Institute has developed adhesives for cementing wires to strain gauges. Some are in commercial production at the Tach'ing Chemical Works and some are in experimental production at the Institute. The adhesives include phenolformaldehyde, polyimides, and silicones. The Institute is also working on structural adhesives and has developed over 20 for commercial use in automobiles, agriculture, and oil field drill bits. Structural adhesives developed include epoxies, silicones, nylons, and polyimides. One interesting use is in the cementing of alloys to the ends of drill bits used in the Tach'ing Oil Fields. Apparently the alloys cannot be welded to the bit end since the welding temperature ($> 1000°C$) destroys the useful alloy properties, so structural adhesives developed at the Institute are used instead. The alloy in question is tungsten carbide and the bit is steel.

Academy of the Lanchow Chemical Industry

Some of the accomplishments of this institute that have been commercialized at local chemical plants are:

- Synthesis gas by both methane/ H_2 conversion and heavy oil gasification to produce CO/ H_2 streams for NH_3 synthesis.
- Steam cracking of gas oil followed by extraction of butadiene using acetonitrile.
- Steam cracking of naphtha for use as a source of ethylene.
- Disproportionation of pine tree rosin to produce a feedstock for use in styrene–butadiene rubber manufacture.
- Biochemical treatment of waste water.
- Anticorrosion agents for metals.
- Kinetics and catalysts for oxidative dehydrogenation of butenes to butadiene.
- Recovery of carbon black from waste water used in the manufacture of synthesis gas (strictly an environmental protection project).
- Development of a total organic carbon analyzer in cooperation with the Peking Analytical Instrument Manufacturing Organization. This instrument operates by pyrolyzing organic materials and measuring carbon in the combustion gases chromatographically. It is a first in China and is being produced commercially in Peking.
- Capillary chromatographic methods for measuring the products of oil cracking.

One of the specific projects under study at this institute is the process study of the oxidative dehydrogenation of butene to butadiene using a small flow reactor. The reactor recirculates unconverted feed. The feed gas consisting of butene, oxygen, nitrogen, and steam is fed using a micro pump developed by the Chinese. The temperature and pressure regimes are 340°–400°C and atm pressure respectively. The reactor is made of glass and is a fixed bed containing 1 g of catalyst in a 3-mm layer sandwich between 23 cm^3 of silica beads. The unit operates at a 50/ 1 recycle ratio with a 4.2 liter/ minute flow rate. Gas chromatography is used for product analysis. The objective of the research program is to obtain kinetic parameters on different catalysts that are prepared elsewhere in the institute. Multicomponent molybdates containing mixed oxides are the catalysts under study.

A second project involves an examination of Pd on alumina catalysts by electron microscopy and X-ray diffraction techniques. The electron microscope was used to define Al_2O_3 physical structure as a function of precipitation and calcination variables and the Pd on the Al_2O_3 surface was being analyzed as a function of the reduction procedure that had been used. In some cases the Pd content was too high for analysis and the Al_2O_3 substrate was then extracted away with hot phosphoric acid. Crystal size is measured by X-ray linewidth and particle size by small-angle X-ray scattering. They also carry out X-ray measurements of the crystallization of polyethylene and are in constant communication with the plant concerning product properties. These can be modified as desired by grafting or radiation. They also measure the orientation of polypropylene fibers. By means of an IR spectrometer, the institute studies the molecular structure of polymers and is specifically studying the structure of the locally produced styrene–butadiene rubber. The latter has the following general composition:

Styrene	25%
Butadiene	
cis-1,4	9%
trans-1,4	14%
cis-1,2	52%

X-ray work at this Institute, both wide- and small-angle, seemed to be of high quality and is carried out by Ho Yi using (Japanese) Shimatzu equipment. Studies are in part concerned with catalyst characterization, particularly with Pd on an Al_2O_3 support. Amounts and crystal sizes of Pd are measured and crystal-size distributions determined from wide-angle peak shape and by analysis of small-angle X-ray scattering peak shape according to a technique proposed by Neilson. Surface areas are measured and compared with those obtained from BET isotherms. Polymer X-ray studies include measurements of the degree of orientation of polypropylene fibers and correlation with modulus and strength. Calculations of Hermans' orientation functions from X-ray data have not been done. Degrees of crystallinity of polyethylene are determined from Hosemann analysis of the dependence of diffracted intensity upon Bragg angle. Intensity changes on heating drawn butyl rubber are observed as a means of following the decrease in crystallinity. A unique technique used in this laboratory is X-ray thermoluminescence. Samples at low temperature (−150°C) are irradiated with X-rays. Upon heating, X-ray emission is monitored. Peaks in emission are found to correspond to mechanical transitions due to T_g and T_m. This method is used to study random and block copolymers of polyethylene with butadiene or styrene. Single peaks are found for random copolymers, but two peaks are found for block copolymers.

Infrared studies are conducted using a (Japanese) Jasco DS79G grating spectrometer in the range of 200 cm^{-1}–1000 cm^{-1}. This is used for organic compound structural studies and polymer microstructure determination, for example, for butadiene-styrene copolymers. No dichroism studies have been carried out. It was indicated that Fourier transform IR spectrometers and Raman spectrometers are rare.

Transmission electron microscopy is carried out by Miss Hu Chun-mei using a (Japanese) JEM7 microscope acquired in 1965 having a 4.5 Å resolution and 250,000 magnification. Studies are made on alumina supported Pd catalysts for which crystal structure and morphology are correlated with preparation conditions. Selected area electron diffraction is employed to identify Al_2O_3 crystal forms. Particle sizes are measured and correlated with X-ray measurements. Polymer studies are on osmium-stained styrene–butadiene block copolymers. Domain sizes are correlated with composition. Attempts are made to characterize domain boundary thickness. Studies are also in progress on particle-size distribution of polystyrene bands and of polybutadiene emulsions.

Lanchow Institute of Petroleum Research

The Lanchow Institute of Petroleum Research is principally engaged in research on lubricating oil additives, catalytic cracking catalysts, hydrofinishing catalysts, new oil products, and new refining processes. The main problems arise from production technology problems, but they also work on technical items that fit into their long-range development plans or that are assigned by the state. For the past 10 years they have cooperated with schools, colleges, and other institutes to develop and manufacture metal organic salt detergents/ dispersants, oxidation/ corrosion inhibitors, extreme pressure lubricating additives and rust additives. One detergent-type additive being produced is polybutene succinimide of molecular weight about 1000. Polyisobutylene is also being polymerized to a molecular weight of several thousand for use as a viscosity index improver for lubricating oils. They also develop and manufacture bead catalysts for TCC units and microspheroidal catalysts for fluid bed catalytic crackers used at other refineries and catalysts used in lubricating oil hydrotreating. They have developed a Dill Chill Process based on a U.S. journal article (Exxon process). The work is essentially applied with a very small basic research component.

The projects specifically shown were involved in examining polybutene succinimides as detergent dispersants. This material was developed in the 60s, but they have sufficient blends for different uses. They are also examining primary amines as ashless additives for lubricating oils.

Institute of Organic Chemistry, Shanghai

While the bulk of the projects are heavily plant-oriented, we were exposed to a project at the Institute of Organic Chemistry that was certainly long-range. This institute has an ongoing project devoted to the production of petroprotein from yeast. This technology has been examined by many U.S. and Western oil companies, and several large plants have been built but are not operating. (There are several large plants in operation in the Soviet Union.)

The Chinese have been doing research in petroprotein for the past 6–7 years and claim to be piloting 2 different processes based on yeast. In one process a gas oil is used as feed and the product is both petroprotein and a dewaxed oil suitable for use as a transformer oil in North China. This process is claimed to be in a pilot plant stage of several thousand tons/year in a suburb of Shanghai.

The second process is based on the use of normal paraffins as a feedstock. The process involves heating (for sterilization) the hydrocarbon feed and added nutrients; passage of feed through a micropore filter into an air-lift fermentor having a residence time of five hours at a pH of 3.5–4.0. The product of the first fermentor goes to a second fermentor to lower residual hydrocarbons and then to a gas separator. The product is water washed several times and centrifuged. Centrifuged material is treated with 0.5 wt% NaOH for nucleic acid removal without destruction of the cell walls. This results in 60%–70% nucleic acid reduction. The petroprotein proceeds to a thin film evaporator and then to drum drier. Trace hydrocarbons are removed by hexane or alcohol (50 ppm level) and the protein is filtered (probably by using spray drying). The nucleic acid products are then film evaporated and drum dried. They are used for agricultural purposes such as plant growth regulation.

Currently the petroprotein is undergoing tests as an animal feed supplement. It is being fed to pigs and chickens in partial replacement of fish meal in their diet. The institute has a joint program with the Shanghai Agriculture Science and Animal Husbandry Institute that involves among other aspects an examination of long-term genetic effects. None have been seen so far in three generations of pigs.

The economic driving force for this project was stated to be an increase in the cost of Peruvian fish meal, which is used as a supplement to supply the necessary amino acid balance in admixture with cereal grains. The major problem still unresolved is the nucleic acid by-product. Nucleic acids constitute 6%–7% of the biomass on a dry basis and the Institute is searching for economic uses. The appearance of *n*-paraffins in the fat of the animals has not been examined.

B. Catalysis

The heterogeneous catalysis effort is highly applied. It essentially concerns (1) process variable studies on existing catalytic systems, (2) the improvement of catalysts for existing processes, and (3) the development of catalysts for processes that exist outside of China but that the Chinese wish to commercialize themselves.

Examples of categories 1 and 2 include numerous studies at various institutes on butadiene production via butene oxydehydrogenation, and oxidation of 1,2,3-trimethyl-4-isopropyl benzene to pyromellitic dianhydride using a V_2O_5/ silicon carbide catalyst (Harbin Petroleum Institute) with the ultimate objective of using Tach'ing bottoms rather than coal tar as a source material. These are essentially process variable studies with microreactor and small-flow reactor equipment using standard catalyst systems. Catalytic studies involve the addition of metal oxides to the base system and a gross evaluation of yields and products.

Examples of category 3 include the development of microspheroidal fluid bed zeolitic cracking catalysts and fixed bed platinum and multimetallic reforming catalysts. There is a considerable knowledge of the Western literature on these subject areas, including the Western patent literature.

China has an abundance of rare earths, which has led to several programs at different institutions aimed at utilizing rare earths as catalysts. At the Lanchow Institute of Petroleum Research we were told, for example, that they had about 10 people working in this field. It was not possible to learn the details, but our perception is that the institutes are screening various rare earths alone and as additives to existing catalyst systems to learn what reactions they will catalyze.

The standard heterogeneous catalyst surface characterization tools were evident at the different Institutes: BET, mercury porosimeters, X-ray diffraction. For example, at the Lanchow Institute of Chemical Industry Pd on alumina catalysts were being examined by scanning electron microscopy (SEM) and X-ray diffraction techniques. However, advanced tools such as ESCA, Mössbauer, Controlled Atmosphere Electron Microscopy, LEED, Auger, and EXAFS were not used. Also, there appeared to be little, if any, solid state chemistry devoted to the synthesis of new materials with potential catalytic value with the single exception of the interest in rare earths. Detailed reaction kinetics involving model compounds also was not in evidence and the Chinese stated that they had only very limited effort devoted to mathematical modeling of catalytic reactions.

Kirin Institute of Applied Chemistry

The catalysis laboratory at this institute is engaged in structural analysis and characterization of heterogeneous catalysts. In this area they do surface area and porosity studies and high-pressure porosity studies as well as BET measurements, X-ray diffractometry, differential thermal analysis, and electron spin resonance, but no Mössbauer, LEED, Auger, or ESCA.

One area of interest is the reaction of ammonia with propylene to form acrylonitrile. At Shanghai a so-called catalyst 49 is being used, probably the Sohio catalyst, which contains P, Mo, Bi, Ni, Co, Fe, and Ca. They are attempting at the Kirin Institute of Applied Chemistry to substitute rare earths for nickel and for cobalt in the hope of improving the process. Acrylonitrile is also made in Tach'ing using a catalyst developed by Dr. Wu Hsueh-chou and his group, a type of P, Mo, Bi catalyst. The conversion of this catalyst is lower than the Sohio material. Another area of interest is the production of methanol from water gas at low pressure ($\approx$ 50 atm) using the ICI methanol process involving copper on alumina and chromia. The Chinese are attempting to reproduce this catalyst system from information available in the literature. The current process that is being used involves a high pressure of 350 atmosphere, 300°C, and a catalyst of ZnO on chromia. Dr. Wu indicated that the hydrogen in carbon monoxide is made by the reaction of water with coal by the Kirin Chemical Company. Also of interest is the oxidation of methanol to formaldehyde using an FeMo oxide on silica in a fluidized bed arrangement. This process is not yet commercial because of problems with the fluidized bed.

Talien Institute of Chemical Physics

Catalytic reforming reactions are being studied at this institute. In the past they have studied platinum reforming catalysts and are now studying multicomponent reforming catalysts. The catalysts under investigation are the bimetallic catalysts Pt–Au and Pt–Ag. They are particularly interested in the study of the reaction mechanism of hydrogenolysis and isomerization. They postulate the cyclopropyl radical as an intermediate for both these types of reactions. This hypothesis is in agreement with their molecular orbital calculations that they have carried out on a Japanese digital integrator.

One project in the heterogeneous area is connected with catalyst evaluation and catalyst kinetics. They are studying the dehydrogenation of long chain paraffins by bimetallic catalysts containing platinum. They are also studying hydroforming reactions. The catalysts include irridium, rhenium, lead, and zinc with platinum, supported on gamma alumina. These catalysts are characterized by X-ray studies, electron microscopy, and hydrogen-chemisorption. For dehydrogenation they find that the particle radius should be greater than 200 Å. They have been successful in developing a reforming catalyst. This catalyst gives 55% aromatics from Tach'ing oil compared with conventional catalysts, which give about 45% aromatics. The instrumentation they have includes small-angle X-ray scattering (Japanese), surface area measurement by BET and by the flow method, and an electron microscope manufactured in Japan. The latter has a magnification of 800,000 and resolution of 70 Å.

Another project deals with the development of new catalysts for nitrogen fixation. The plan is to make nitrogen complexes to activate the nitrogen. The catalysts used are electron donor–acceptor complexes. They have achieved reaction of nitrogen with hydrogen at 350°C. They started initially with the Japanese work (Tamaro) that makes use of graphite containing $FeCl_3$–K. The latter catalyst has a low activity. Research workers at the Talien Institute have shown that the potassium is catalytic and is not a reducing agent. The iron in the catalyst is reduced to Fe^0. They have developed new supports for the catalyst, namely Al_2O_3 and activated carbon. These catalysts have a better activity than the Tamaro catalysts. The catalytic activity depends on the content of the $FeCl_3$. They have also used ferrocene as a source of the iron. These catalysts have a much higher activity than commercial synthetic ammonia catalysts. However, the stability of the catalysts is unsatisfactory. The degradation of the catalyst is associated with the growth and the size of the iron crystallites. The plans are to study different iron complexes using large molecular ligands as complexing agents and the interaction of the catalyst with the support.

The high vacuum equipment in this laboratory was outstanding. They had Kovar-to-glass seals and excellent flexible metal bellows to join glass to metal systems. They had Varian high vacuum valves and flanges that are imported from Varian, Canada, and they had Balzer high-temperature ultra high vacuum valves.

Futan University

The catalysis group at Futan University does applied work through a connection with the Shanghai Scientific and Technical Committee. This involves industrially sponsored work (probably some originating from the Shanghai Petrochemical Complex) and is organized by Professor Kao Tsu. Much of the applied work involves the screening and development of catalysts. Studies include Ziegler-Natta $TiCl_3$–Al alkyl catalysts for polypropylene polmerization. Batch polymerizations are carried out in 2-liter, 30-liter, and 1300-liter autoclaves. (The latter is a very large apparatus for a university laboratory and costs 100,000 yuan.) Activities of 15,000 g/g with 95% tacticity have been achieved. Current efforts are devoted to development of a solvent-free, solid-phase catalyst system that would not require washing residual catalyst from the polymer. The approach is to complex the catalyst with an organophosphorus compound. The chemistry department has an association with a small scale plant in Shanghai that is producing polypropylene sweaters and socks and (wood-free) pencils.

The catalysis unit is in a separate building and is funded by the Shanghai Scientific and Technical Committee so that the various polymer factories in the Shanghai area may make use of its instrumentation and expertise. The building is exceedingly well-instrumented. It includes a Rigaku rotating anode, X-ray powder diffractometer, and low-angle camera. It also includes a large number of high-pressure liquid chromatographic units, a Mettler T160N balance, and a large number of semiworks scale autoclaves. The lab devotes part of its time to helping local industry and part of its time to the training of students.

Shanghai Institute of Chemical Engineering

This institute does quite a bit of catalytic research including zeolite catalytic cracking and the oxidation of ethylene over silver catalysts. It is not clear whether these catalytic experiments are meant to be undergraduate teaching laboratories to repeat past experiments or whether they are graduate research. We also saw a molecular sieve dewaxing system with a huge 6-tower pilot plant. Normal paraffin is absorbed on molecular sieves for the production of petroprotein. This apparatus cost 100,000 yuan for the equipment and parts, and required over one year to build. We also saw a project of dehydrogenation of ethyl benzene to styrene over ferric oxide catalyst of 10–12 mesh size. This is said to be a third-year student research laboratory. We also saw a distillation column with five glass sieve plates to study point efficiency. Data is computer collected for the study of temperature and gas velocities on each plate. To study the concentrations, the samples must be removed by hand and titrated. The unit operation laboratories also include studies of friction factors, Reynolds numbers, centrifugal pumps, and heat exchangers. There was study underway of the dynamics of a distillation column for ethanol and water. There is no computer to receive the data, so the results are recorded on charts. They also have in the unit operation laboratories extractors, absorption towers with Raschig rings, and rotary filters.

Institute of Chemical Physics, Lanchow

An area of activity at the Lanchow Institute of Chemical Physics was Oxo Synthesis Catalyst Preparation. The work involves the synthesis by base-transfer catalyzed reaction of $PhCCo_3(CO)_9$, and the study of the physical properties of this material. They also outlined a synthesis for the complex of $RhCl(CO)(PR_3)_2$ in which each of the phosphines is attached to a silicone backbone. They have thus synthesized a phosphine containing silicone and have made a transition metal complex of it. They indicated that they hoped this would be active in the oxo reaction. We pointed out that the active species in the oxo reaction catalyzed by rhodium was probably $RhH(CO)(PR_3)_2$; they appeared to be aware of this but indicated their Cl compound would probably be converted to the hydrido compound under reaction conditions.

Another area of work at this institute is concerned with oxidative dehydrogenation catalysts with which they were converting butenes into butadienes. This work appeared to encompass a set of five or six different catalysts including product analysis and evaluation of reaction rates. The catalyst preparations are apparently done at this Institute; the performance as a function of temperature and pressure is determined in a relatively large testing program.

A third area of work involved a temperature programmed desorption apparatus in which they are using various solid catalysts for gas uptake. Incidentally, this work involves some vacuum line systems that contain glass stopcocks but no Teflon stopcocks. We were shown the gas chromatography–mass spectrometer system with associated computer built in China. This is a prototype instrument built in Peking. A final area involved catalytic kinetics and was concerned with the oxidation of *n*-butenes over Bi, Mo, P catalysts.

C. Polymers and Synthetic Scientific Fibers

Synthetic polymer work in China includes efforts to synthesize polymers by the usual radical and ionic routes, to make block copolymers by the living polymer technique, and to prepare condensation polymers of polyesters, polyamides, and polyimides. The strongest efforts (described in Section B) are to find new polymerization catalysts. There are also strong efforts to produce specialty polymers having unique functions such as the crown-ether containing polymers for analytical separations, blood plasma substitutes, and polymers with functional groups having pharmacological activity.

Molecular characterization of polymers is through the usual spectroscopic techniques of infrared and NMR primarily using imported instrumentation. High-field (superconducting magnet) NMRs are scarce (one seen) as are ^{13}C Fourier transform NMRs. Fourier transform IR spectrometers and laser Raman spectrometers are also rare or nonexistent. Digitalization of instrumentation is not common. Molecular weights are generally determined using viscosity, imported vapor phase osmometers, homemade membrane osmometers, and homemade or imported light scattering photometers. No laser solution light scattering studies involving linewidth measurements were seen. Gel permeation chromatographs (GPC) of the imported (Waters) type were not apparent; but an appreciable effort has gone into making homemade GPCs shown, for example, in Fig. 15 which are just beginning to be commercialized in China. Strong emphasis was placed on the development of packings for GPC columns.

Solid-state studies involving X-ray diffraction and scattering were undertaken usually using imported equipment. These involve

Figure 15. Chinese-made Gel Permeation Chromatograph (GPC).

degree of crystallinity, orientation, or crystal or particle size measurements. Little work on polymer crystal structure was evident. Solid-state infrared is generally undertaken for analytical purposes, but there is some use of dichroism for orientation studies. Solid-state NMR studies were not common, and no pulsed NMR studies on solids were encountered. Laser small-angle light scattering studies (photographic) were in progress in two laboratories, as shown in Fig. 16, both for studies of spherulitic polymers and liquid crystal type studies. Electron microscopy was in use in principal laboratories. Transmission electron microscopes were usually Japanese, but Shanghai-made scanning microscopes were in use, as shown in Fig. 17. No scanning transmission electron microscopes (STEM) were encountered. Surface studies by techniques such as ESCA (electron spectroscopy for chemical analysis) were not encountered. It would seem that these should be important in view of the strong interest in catalysis. Mechanical property studies, other than routine tensile testing, were not extensive. Several laboratories used homemade torsion pendulums and one had a fairly sophisticated digitalized Japanese viscoelastic spectrometer. Conventional dielectric loss equipment was available in several laboratories. All in all, solution- and solid-state characterization procedures were relatively routine and little novel work was seen. Scientists were trying hard to implement techniques used

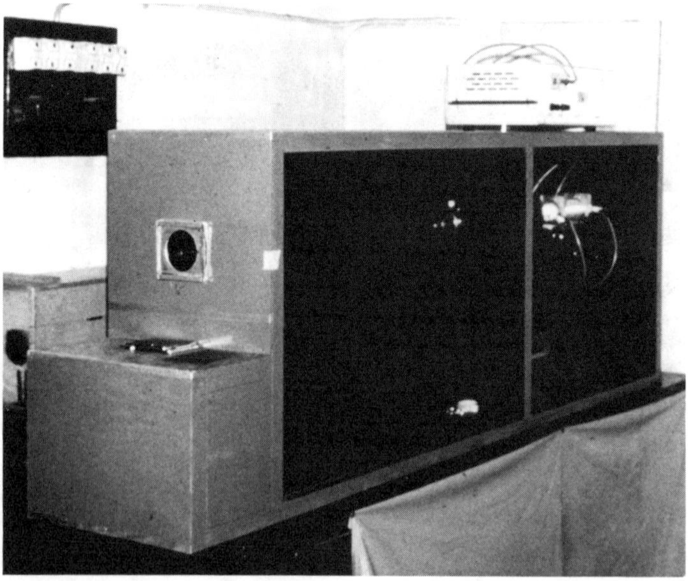

Figure 16. Laser light-scattering apparatus at the Institute of Chemistry, Peking.

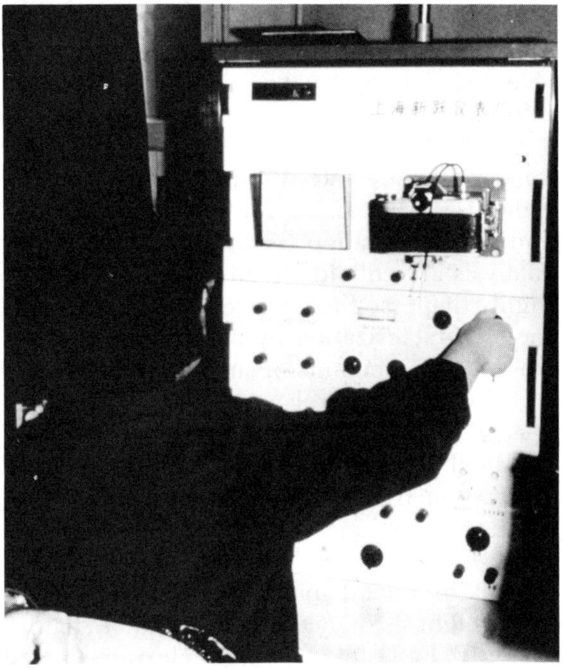

Figure 17. Shanghai-built scanning electron microscope at the Institute of Chemical Industry, Lanchow.

abroad but were often hampered by lack of modern equipment, forcing them to devote extraordinary time to instrument development. They were very conversant with Western developments and are well prepared to make advances with the development or acquisition of suitable techniques.

An area of low activity was that of melt rheology. (It may be that some was in progress in engineering-type institutions that were not visited). No capillary or cone and plate-type equipment was seen, and there did not seem to be great interest in this field. This was somewhat surprising in view of the relevance of this type of work to practical areas of polymer processing.

There also seemed to be little activity in theory. No work on chain configuration or thermodynamics was encountered, nor were studies of chain dynamics or theories of glass transitions or crystallization seen. This is undoubtedly a consequence of the suppression of theoretical efforts during the Cultural Revolution and the emphasis upon practical goals. It would seem that as Chinese polymer science matures, more emphasis must be placed on theoretical goals in order to provide guidance and inspiration for more novel experimental efforts. This will be difficult because of the "missing generation" of scientists who did not receive strong theoretical training during this period.

As in the United States, colloid science does not generally appear as a separate discipline at most universities or institutes, but rather is studied in the context of other fields. Much of the preparation and characterization techniques in the strongly emphasized area of catalysis are colloid chemistry related. These involve preparation of highly disperse high surface area dispersions of active materials and their characterization by microscopy, scattering, and gas adsorption. Appreciable amounts of such work are in progress.

Colloids are also important in forming solid polymers having special properties such as impact resistance. These often consist of colloidal dispersions of a rubbery polymer (size range 0.1–10 μm) in a glassy matrix. The dispersion is often prepared by emulsion polymerization and may involve stabilization using surface active agents or block copolymers. There are limited studies of this type in progress.

Colloidal dispersions that may or may not involve high molecular weight substances are important in paint, food, lubricant (graphite colloids), and fuel (coal or oil slurries). No appreciable work in these areas was encountered, although it is most likely to be done in highly applications-oriented institutes or industrial laboratories that were not visited.

Institute of Chemistry, Peking

There is a very appreciable polymer effort at the Institute of Chemistry. This effort is headed by Professor Ch'ien Jen-yuan (who studied Raman spectroscopy at the University of Wisconsin). The polymer program started in 1953 at the Institute of Organic Chemistry in Shanghai and moved to Peking in 1956. Work started with molecular weight measurements and fractionation (viscosity, osmotic pressure, vapor phase osmometry, light scattering, and ultracentrifugation). In the latter 1950s work was initiated in studying the nature of polymer solutions. Studies of the effect of molecular weight distribution on the second virial coefficient were reported at the Prague IUPAC in 1975 using light scattering and osmotic pressure techniques (light scattering was measured using a homemade apparatus similar to a Brice-Phoenix). Sedimentation rates (using a Spinco ultracentrifuge) were studied in theta solvents. Solvent–nonsolvent fractionation was studied and interpreted on the basis of Flory-Huggins theory, and phase diagrams were established. Mechanisms of ultrasonic degradation of polymers in solution were also studied showing a narrowing of molecular weight distribution. A limiting molecular weight below which degradation did not occur was noted. Work on solutions was suspended during the Cultural Revolution as it was not considered to be sufficiently practical. It is currently being resumed.

Recent solution studies primarily involve use of GPC. One of the specialists in this area was Shih Liang-ho. An instrument was developed at the Institute (judged to be somewhat inferior to a Waters, Associates instrument) and is now in commercial production by an instrument manufacturer. The instrument utilizes both differential refraction and ultraviolet absorption-type detectors. Emphasis has been placed on the development of new column-packing materials. Simulation of elution curves has been carried out. Packing materials include copolymers of divinyl benzene crosslinked polystyrene with ion exchange resins. The column resolution is determined by interaction between groups on the polymer and the substance to be studied. These have been used to study trace H_2O in organic compounds. Highly crosslinked columns with small pore size have been used for steroid and amino-acid analyses.

Solid-state and polymer morphology studies have been primarily conducted by Hsu Mao (who visited the United States in 1977 with the Chinese chemistry delegation). Hsu has built a photographic small-angle light-scattering apparatus (SALS) (following Stein's UMass design) shown in Fig. 16. He has been applying this to the study of liquid crystals of aromatic polyamides in H_2SO_4 solution. He observes a thermotropic nematic to cholesteric transition analogous to what is observed with low molecular weight cholesterol esters. At this transition the H_v light scattering pattern changes from one having cylindrical symmetry to a four-leaf clover type suggesting nonrandom orientation correlations. This transition is also studied by depolarized light intensity techniques. The change in the scattering pattern on deforming the solution is seen upon microscopically examining such solutions by the appearance of a streak-type pattern having a periodicity corresponding to diffraction from striations. The study of this system is obviously of interest from the point of spinning high modulus fibers of the "Kevlar" type.

Hsu also is using the SALS technique for studying crystallization kinetics of polyesters. This is a continuation of earlier work on crystallization from this laboratory by Ch'ien Jen-yuan (*Gaofenzi Tongxun*, **1965** *1*, 264) on the effect of uniaxial stretching on the crystallization process of polyethylene terephthalate fiber. In these dilatometric studies on samples that were drawn at room temperature and then heated, it was concluded that upon drawing to 300% the Avrami exponent decreases from 2.8 to 1.2 and the crystallization rate *decreases*.

Hsu also utilized the SALS technique for the study of crystallization of polyethylene and polypropylene. A noted deficiency in studies of polymer crystallization was the lack of a differential scanning calorimeter (DSC). This apparatus has become one of the most widely used in the United States for determining amounts and rates of crystallization. No homemade or commercial apparatus was available in this laboratory for this purpose. Infrared spectrometry is used for characterization using an old German Zeiss URIO as well as an American Perkin–Elmer Model 180 spectrometer. This is probably one of the best available grating spectrometers. Some of the electronics had been replaced by Chinese components. IR polarizers of the AgBr grid type were available and dichroism studies were in progress. Infrared was used for tacticity determinations of polypropylene, studies of hydrogen-bonding in nylon-66, and in comparing spectra of vinyl acetate-acrylonitrile copolymers and blends. Interactions between cyanide groups on the chain were seen to affect the spectra and scientists hope to learn about sequence distribution from their study. The position of the C≡N stretching frequency was correlated with the pair distribution probability for adjacent acrylonitrile units on the chain. The effect of water on the transition between α and γ forms of nylon-66 was studied. Infrared spectra of fibers were observed on an array of single fibers. No Fourier transform infrared type equipment or Raman spectrometers are available.

Dynamic mechanical spectroscopy was studied using a Japanese instrument manufactured by Uamoto Company. This instrument is similar to a Rheovibon and was automated using Chinese electronics so as to give digital printout of data. This is a fine apparatus that is definitely state-of-the-art. It was used to determine the anisotropy of the viscoelastic spectrum of polyethylene terephthalate produced by stretching. A homemade torsion braid analysis apparatus and a Japanese thermogravimetric analysis apparatus are available.

NMR studies are carried out for characterizing polymer microstructure using a French made RMN-250 250-MHz superconducting NMR apparatus that was inherited from the Institute of Photography. Liquid helium is obtained from the liquefier at the Institute of Physics (liquid N_2 is commercially available but liquid He is not). A Fourier transform ^{13}C apparatus (Varian XL-100) is available at the Institute of Photography (about 10 kilometers away). ESR instrumentation was said to be available but was not seen.*

X-ray studies are carried out on a Phillips four-circle diffractometer at the Institute for Biophysics giving punched tape output.

* A home-built ESR spectrometer is apparently available in the Institute of Chemistry, Peking.

Computations are done on a Chinese made TQ-16 computer with a 32K memory and 10 μsec cycle time.

A notable study was the use of pyrolysis gas chromatography (GC) for defining polymer microstructure. (This is representative of the very fine and extensive GC efforts seen throughout China.) Initial studies were made on an East German instrument but an instrument of comparable quality has been made in China which has served as a prototype for a commercially available model, variations of which were seen in many Chinese laboratories. Much effort was devoted to the development of new column packings. (There exists an instrument development and manufacturing facility associated with the Institute of Chemistry.) This technique was applied to the study of long-chain branching in polymethyl methacrylate, which was compared with model compounds. It was shown that branching occurred at tertiary carbon atoms. Studies were also made on polystyrene crosslinked with divinyl benzene. It was stated that degrees of crosslinking could be determined in 10 minutes on very small samples. The microstructure of polybutadiene was determined by this means more rapidly than by IR. The method was also used for the determination of small amounts of a comonomer in a polymer, for example, 1% per fluorinated propylene in tetrafluoroethylene. 2% of nylon-6 in nylon-66 was detected. Pyrolysis using a CO_2 laser was attempted but abandoned because of lack of reproducibility.

Studies were in progress on block copolymers having polytetrahydrofuran soft segments and polyethylene terephthalate hard segments as possibilities for a thermoplastic rubber useful as a substitute for rubber bands.

Some work was in progress on composites involving polyamides, but the effort did not appear extensive. The only work on inorganic polymers was on silicones, which represented the pioneering work for the silicone industry in China. Some work on doped unsaturated polymers as semiconductors was in progress. Photoconductive polymers for holography were being developed involving charge transfer complexes with polyvinyl carbazole. These polymers work by causing thickness changes when the charged polymer film is heated to above its glass temperature. Theoretical work was described on the use of molecular orbital theory to predict properties of polyenes. Predictions of absorption spectra, ionization potentials, polarographic reduction potentials, NMR chemical shifts, and reactivities were considered for homologous series having different substituent groups at the same chain position or the same groups (say NH_2) at different chain positions.

One area of emphasis is in the use of $TiCl_3$-based catalysts for low-temperature polymerization of polypropylene. Catalysts are characterized by X-ray diffraction, surface area measurements, and electron microscopy. High-activity catalysts with small crystal size (30 μm) are prepared leading to nascent polymer particles of 20–30 μ. These are 96% isotactic and consequently are highly crystalline. There is an effort to correlate polymer microstructure with processability.

There is an effort to develop organic sulfur compounds (e.g., thiobisphenol) as melt stabilizers. Mechanisms of oxidation are being investigated. Attempts are being made to modify polypropylene fiber to improve aging and dyability so as to render it better for clothing purposes. This is done by copolymerization with 5% of monomer units such as

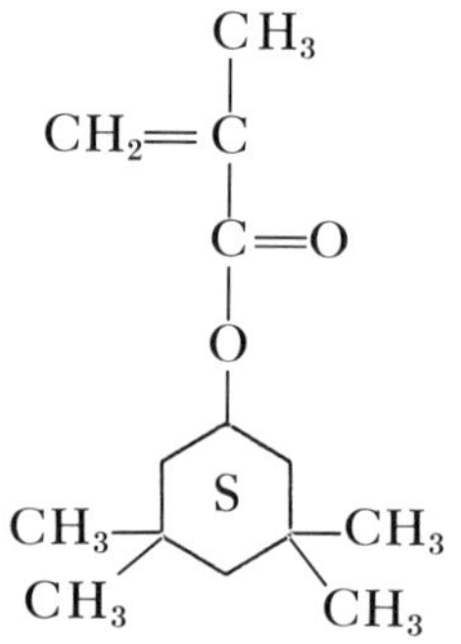

This copolymer is blended with polypropylene.

There are studies on end-capped polyimides with improved solubility. These are intended for wire coating. There are also studies of polymerization of dinitriles and of polymers containing triazine rings in an effort to produce polymers of improved temperature and thermal deformation stability. Such dinitrile polymers are typically

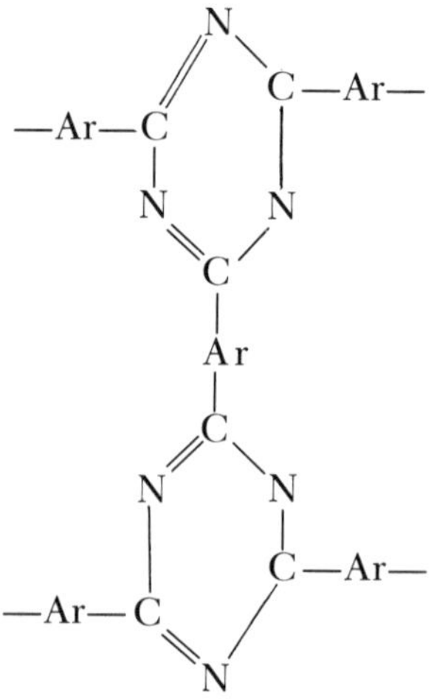

Peking University

Polymer research at Peking University is led by Professor Feng Hsin-te, who received a Ph.D. from Notre Dame in 1945 working with C. Price. An area of emphasis in the polymer laboratory is the synthesis of polymers of biomedical interest. An example is

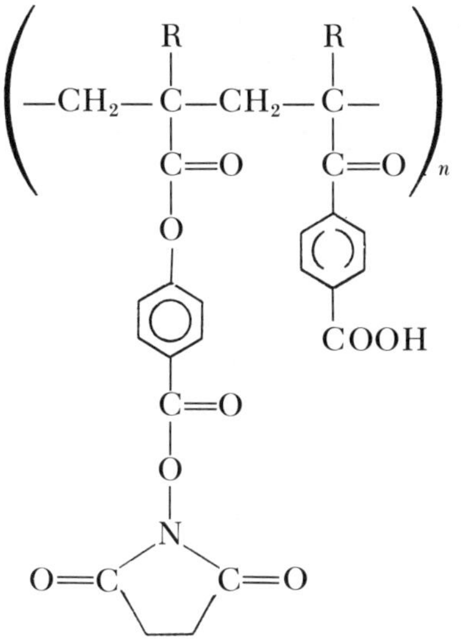

This polymer reacts with proteins, enzymes, etc., and is used as an absorbant to react with blood serum, antibodies, etc. It is said to be comparable to or better than Sepharose 4B.

There is also a program to study Ziegler-Natta type catalysts for polypropylene polymerization. Studies are in progress to examine the α, β, and δ forms of $TiCl_3$ catalysts using electron microscopy. The transformation of crystal forms from β to δ in the electron beam has been observed. Electron microscopy is carried out by a scientist who was educated in the Soviet Union under Kargin and who uses a Japanese JEM-60 electron microscope.

Institute of Chemical Physics, Talien

Catalyst development for polypropylene polymerization is under study at the Institute for Chemical Physics in Talien. High efficiency Ti-based catalysts for high-pressure (30 atm) polymerization yield 40,000–50,000 g polymer/g Ti. With supported catalysts 100,000 g polymer/g Ti has been obtained but with somewhat lower tacticity. The activity of the supported catalyst is greater than that of $TiCl_3$.

Catalyst systems were studied with ESR by which spin concentrations were correlated with catalyst activity. Molecular weights were regulated with H_2, but this was not possible with Cr_2O_3 on silica. Research is also underway on ethylene–propylene copolymers of relatively low molecular weight to be used as viscosity index improvers for lubricating oils. There is an effort to make single-component catalyst systems to eliminate aluminum alkyls. There are plans to investigate the catalyst activity of rare earth *f*-orbital catalysts.

Techniques for characterizing the polymers produced are relatively unsophisticated. Molecular weights are characterized by intrinsic viscosity. No average molecular weight or molecular weight distribution studies are in progress. A crude homemade gel permeation chromatograph is available. Polypropylene tacticity is characterized by extraction and by infrared spectroscopy. Catalysts are physically characterized by small angle X-ray scattering, carried out on a (Japanese) Shimadzu apparatus and by electron microscopy. Microscopy is carried out using a Chinese-made scanning electron microscope (SEM) having 800,000 × magnification with 70 Å resolution equipped for photoelectron emission studies for elementary analysis. Surface areas are measured by BET isotherms and by flow methods. No surface chemistry studies by ESCA or related techniques are underway but there is hope of instituting these in the near future.

There is a polymer-related project of significance to bioanalytical chemistry. The project involves preparing polymers of crown ethers of the sort

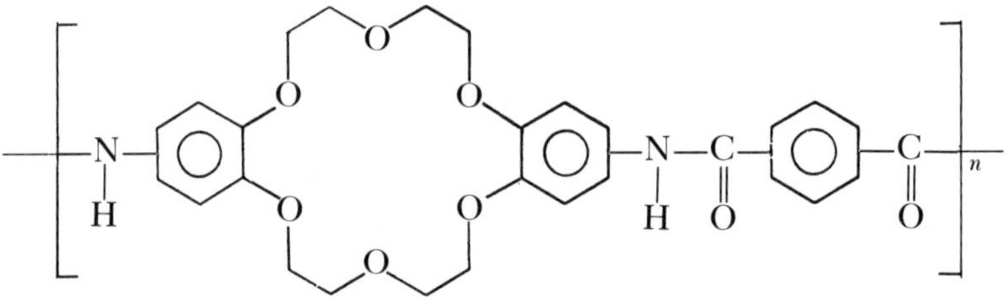

with molecular weights of the order of 100,000. Solubility properties of these have not yet been determined. It is proposed that these be used for analytical separations.

Kirin Institute of Applied Chemistry

One of the polymer efforts in a group led by Professor Ch'ien Paokung is concerned with the preparation and characterization of *cis*-1,4-polybutadiene. This polymer, prepared with a nickel catalyst, has been characterized by studies of microstructure, morphology, gel content, intrinsic viscosity, molecular weight, distribution, branching,

Mooney viscosity, cold flow index, density, stress–strain curve, and stress relaxation. The polymer prepared at the Kirin Institute has a gel content of less than 1% and a molecular weight distribution broader than polymers obtained abroad. The scientists here feel that Mooney viscosity is not adequate for characterizing processability and that it is desirable to know stress–relaxation properties. They believe that differences in processing properties are related to branching. These correlate yield strength with intrinsic viscosity for this polymer as well as for natural rubber and polyisoprene. Activation volumes are obtained by Eyring analysis of stress–strain data. Activation for yield has been obtained to interpret morphology in terms of Yeh-type nodular structures (20–30 Å), aggregates (100–1000 Å) and gel.

Another study in this group by Chiang Ping-cheng deals with the effect of structure on the stress cracking of FEP copolymer (C_2F_4–C_3F_6). Optical and electron micrographs (Hitachi SEM) and small-angle light-scattering studies have been conducted. The latter were made with a homemade instrument involving a small Chinese He-Ne laser. Chinese-made polarizers are apparently available. Spherulitic (0°–90°) type H_v patterns are obtained at high rates of cooling and rod-like patterns at low rates of cooling. These patterns are used to measure spherulite size for spherulitic samples. SEM studies have been made of fracture surfaces and the size of the fracture pattern is not correlated with spherulite size. Failure is related to flaw density.

This laboratory is equipped with a dielectric loss apparatus (homemade), a homemade torsion pendulum and a homemade GPC. There is no equipment for melt rheology. A homemade DSC is under construction and preliminary scans are available of the melting of low-density polyethylene (these seem to be equivalent to what might be obtained with a Perkin Elmer DSC-1).

A study in the polymer chemistry group by Wang Fu-sung is concerned with polymerization of isoprene and butadiene using rare earth catalysts. Associated with this effort is the production of synthetic zeolytes to serve as catalyst supports. The variation of catalytic activity among the 14 rare earth elements was studied and found to differ markedly (Sm and Eu have exceptionally low activity). This variation does not correlate with the valence state nor with charges in the IR or Raman spectra of the metal complex with tributyl phosphate, presumably due to differences in *f*-orbital bonding (sp^2f^3 hybridization). The energy differences between the complexes of the various elements are correlated with the activity. The microstructure of the polybutadiene is found to differ slightly ranging from 94% *cis*-1,4 to 96% in going from the lighter to the heavier rare earths. In a manner similar to alkyl-Li catalysts, the molecular weight distribution is found to shift toward higher values. It was demonstrated that rare earth catalysts can give rise to "living polymers" and that isoprene–butadiene diblock copolymers can be prepared by this route. These block copolymers were compared with blends of polyisoprene with polybutadiene and found to give greater elongation (1000% vs 300%), tensile strength, and green strength. (The role of crystallinity and microdomain formation in affecting these properties was not considered.)

Institute of Organic Chemistry, Shanghai

The polymer work at the Institute of Organic Chemistry in Shanghai falls into two categories: (a) natural polymers and (b) synthetic fluoropolymers. In the former category are studies on blood plasma substitutes such as carboxy methyl amylose prepared from cornstarch by T'u Ch'uan-chung, somewhat reminiscent of the work on "Dextran" in the United States some 20 years ago. Solutions exhibit typical polyelectrolyte behavior. Samples are characterized by paper chromatography using ^{14}C tracer and by molecular weight distribution studies involving fractionation and osmotic pressure measurements. Enzymatic degradation rates are found to be appreciably suppressed with increasing degree of substitution. Synthetic plasmas have been animal tested and are now commercially available for clinical use.

A unique technique featured at this laboratory by Wang Yung-lu is CO_2 laser-induced pyrolysis gas chromatography. This method, judged not reproducible by the Institute of Chemistry in Peking, is employed at Shanghai for polymer analysis. Initial work was carried out using a pulsed ruby laser, but it was necessary to add graphite to transparent samples for there to be sufficient absorption of radiation. An advantage of the CO_2 laser is that its output is in the infrared (10.6 μm), where most polymers absorb. Consequently, small samples at the focus of the laser pulse are quickly heated to ~7000°C and then rapidly cooled to reduce the probability of complicating secondary reactions. N_2 carrier gas conveys decomposition products through the gas chromatograph. It was claimed that it is possible to determine sequence distribution in block copolymers and degree of branch in 3–5 minutes by this technique. There are plans to commercialize this instrument.

Polymer physics and chemistry are carried out in Dr. Huang's group under the direction of Shih Kuan-yi (Physics) and Tai Hsing-yi (Chemistry). Major research equipment includes a homemade DSC and TGA, a Japanese (Shimatzu) DTA, a Japanese (Shimatzu 1s 500) tensile tester, a Japanese (TR 10C) dielectric bridge, a German (Knauer) vapor pressure osmometer and a French (Sophica) light scattering photometer.

Following Huang's interests in fluorine chemistry, major efforts are concerned with fluoropolymers including FEP copolymer (C_2F_4–C_3F_6) and Fluoroplastic 40, a EPTE (C_2H_4–C_2F_4) copolymer. These are of interest for corrosion-resistant metal coatings and for wire insulation. Work on fluoropolymers was initiated at the Peking Institute of Chemistry but transferred to Shanghai because of the proximity of related industry (a Freon plant). Fluoropolymers are widely used and are principally manufactured in a factory in the Shanghai area. The fluoropolymers are characterized by electron microscopy (a Shanghai-made SEM, 100 Å resolution). Correlations are made between spherulite morphology, composition, and stress cracking. Compositions are established by IR, laser pyrolysis, DSC, and DTA. Melting points are found to be linear functions of the composition of the copolymer. The heat of fusion is found to be dependent upon composition, suggesting the inclusion of CF_3 in the crystals as point defects. Effects of cooling rate and annealing on this phenomena are being studied. Crystallization kinetics are studied by DSC. The Chinese believe that fractionation occurs upon crystallization from the melt. X-ray analysis of crystal imperfection has not been done but is planned.

While organoelemental chemistry (involving elements in addition to carbon) is an area for study at this Institute, no studies on inorganic polymers (e.g., polyphosphazines) or silicones are in progress. Silicone studies are said to be done in a university in Canton (not visited) and at a laboratory of the Ministry of Chemical Industry. Silicone rubbers and coupling agents are said to be commercially produced.

Futan University

The polymer group at Futan University is led by Professor Yü T'ung-yin, who is an organic chemistry graduate from the University of Michigan. About one-third of the chemistry students major in polymer science, but this is decreasing (to about one-fifth) as other areas of specialization develop. One of the projects by Hsu Ling-yun involves studies of SBS (styrene–butadiene–styrene) type thermoplastic elastomers. These are synthesized by sequential monomer addition to a living polymer and have styrene segments with molecular weight of 10,000 and butadiene segments (90% *cis*-1,4) of molecular weight 80,000. ABS (acrylonitrile–butadiene–styrene) polymers are being studied by Chu Wen-hsuan. He is also preparing high impact polystyrene using rubber polymerized in suspension to give small particles. Morphology is studied using an electron microscope of Japanese manufacture in the Biology Department. They are studying the relationship of particle size to shear bonding and crazing and conclude that best results are obtained with a mixture of oil and 2 μm rubber particles in about equal amounts prepared by mixing samples made my emulsion and suspension polymerization.

Shanghai Institute of Chemical Engineering

The polymer program at this Institute is directed by Professor Li Shih-chin (who wrote the polymer physics book used in teaching), Associate Professor Chao Te-jen (who studied at the University of Manchester with G. Allen), and Lecturers Wu Ho-jung and Lou K.-t. Their equipment includes a Chinese-made X-ray diffractometer (Geiger counter scanning), which is currently used for catalyst studies but will be used for polymers. One polymer project under study is particle-size distribution measurement in polyvinyl chloride emulsions using a centrifuge in which the distribution is scanned and using a photodiode with digital output.

The polymer lab contains instruments for DSC and DTA, as well as an X-ray diffractometer, an instrument for determining particle size distribution for polyvinyl chloride, and a centrifuge with the capability for photoelectric measurement of concentration gradients. The laboratory is adequate for undergraduate teaching but not for graduate research.

Students in a synthesis teaching laboratory were studying ring opening polymerization of epichlorhydrin and of polypentene. This course, now elected by third-year students, has them spend 10 hours per day for 5 days per week in this laboratory for a semester with the sixth day at course work! The second semester is spent in a polymer physics lab. About 8% of the chemical engineering students elect to concentrate in polymer studies in their last year. A fourth-year program now being planned will involve a one-semester full-time research project. Plans for the graduate program are still uncertain.

Institute of Chemical Physics, Lanchow

Four polymer-related problems were mentioned at this institute. One of these deals with the preparation of polymer membranes for selective gas permeability, a second with the preparation of polymeric enzymes, a third with the construction of a K^+ selective electrode formed by surrounding a Ag/ AgCl/ KCl electrode with a polyvinylchloride membrane containing a crown ether (formed by solvent casting from a mutual solvent), and a fourth with the preparation of crown ether based polymers of structure

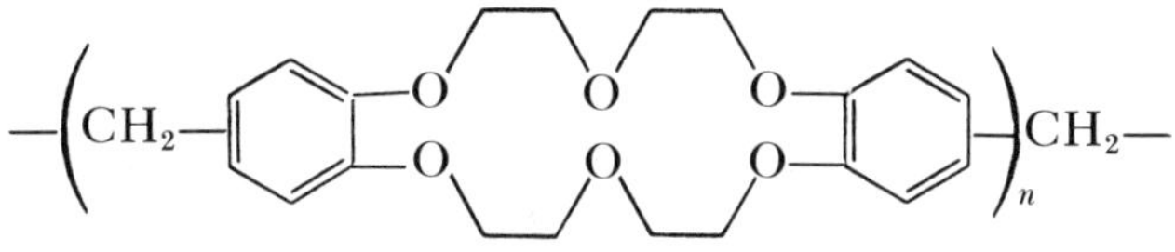

for separations of inorganic ions. (This work is somewhat similar to that previously described in Talien.) There is some hope that these might have value for rare earth separations.

D. Laser Chemistry

Laser chemistry is an area China will be pursuing quite vigorously in the near future, although the current level of activity is still low. Many institutions have either just started or plan to start some programs in the area of laser chemistry.

Laser units have not been mass produced in China except for small He–He lasers. Consequently the availability of lasers and associated equipment to chemical laboratories is still very limited. All the laboratories that have been engaging in laser chemistry research have had to construct their own lasers before they could initiate any research projects, a major handicap for developing laser chemistry programs. In many of the laboratories we visited, it was evident that the ability of Chinese experimentalists to get lasers running is excellent. Many scientists seem to be knowledgeable, and they are not afraid to tackle difficult tasks. At present most of the laboratories are in the stage of setting up laser equipment rather than doing active research on laser chemistry problems.

It is important to note that "Laser Industry" is one of the eight priority items in the current drive for the modernization of science and technology as stated in the report delivered by Vice-Premier Fang Yi at the National Science Conference held in March 1978. Along with the development of the laser industry, it is expected that China will make progress in the area of laser chemistry in the near future.

Among the institutions we visited, the following have current laser chemistry-related programs.

Institute of Physics, Peking

A CO_2 TEA laser is in use with an output of several joules/ pulse, and multiphoton dissociation of BCl_3 is under investigation. The observation and analysis is entirely based on the chemiluminescence. The spectra of instantaneous and delayed fluorescence are being analyzed in an attempt to understand the dissociation mechanisms. This work repeats the work carried out by Rockwood and Rabideau at Los Alamos in 1973.

Kirin Institute of Applied Chemistry, Ch'angch'un

A relatively short, cylindrically shaped, metal-enclosed CO_2 TEA laser shown in Fig. 18 with an output of 300 mjoule/ sec (line-tuned) is in use to study the multiphoton dissociation of CF_3I for the separation of ^{13}C. This program seems to be quite successful. They have been able to enrich ^{13}C as Paul Houston did about a year ago at Cornell.

A homemade Ar-ion laser with an output of 7 W was used to pump a dye-laser (jet arrangement). Using homemade Rhodamine-6, their conversion efficiency was reported to be 10% to about 15%. Spectroscopy of highly excited atoms, two-photon absorption, multiple-photon absorption, opto-acoustic detection, and laser Raman spectroscopy are the types of work they will pursue in the near future. They have just begun operation of the dye laser system, and we were surprised to find the laser beam bouncing around in front of us without any safety precautions, as shown in Fig. 19.

Figure 18. CO_2 TEA laser at the Kirin Institute of Applied Chemistry.

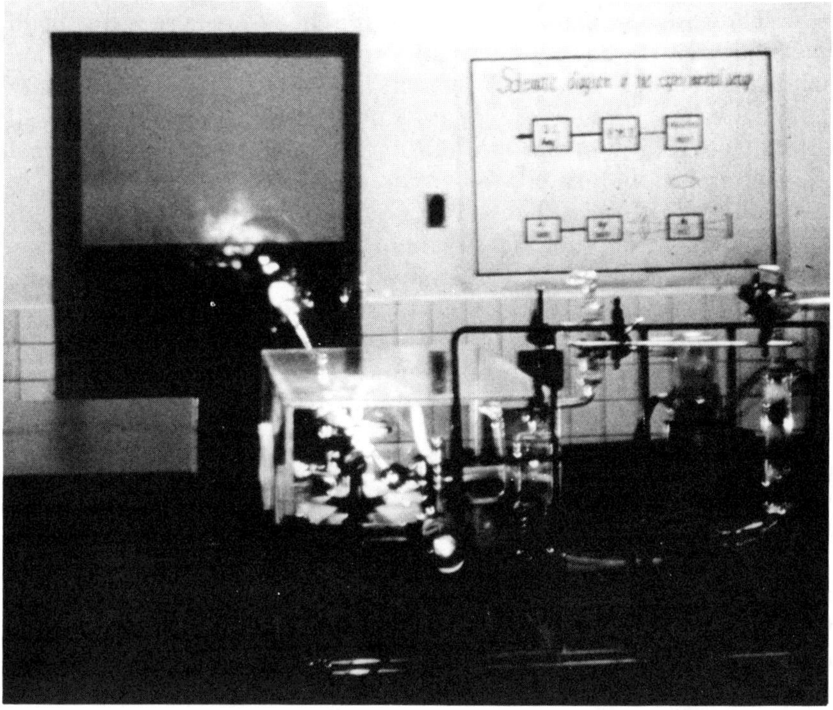

Figure 19. Ar-ion laser being used to pump a dye laser at the Kirin Institute of Applied Chemistry.

Futan University, Shanghai

Faculty members of the Chemistry Department are planning to initiate a laser chemistry program in the near future in collaboration with the members of the Department of Optics. In the Department of Optics we saw several lasers:

> Sub-nanosecond CO_2 TEA laser system under testing. This laser system follows the design of Yablonovitch (Harvard). A sub-nanosecond (≈ 500 pscc) system will be used for the study of the multiphoton excitation and dissociation of polyatomic molecules. The TEA laser used in this system is a spirally arranged pin electrode laser.
>
> Nitrogen laser pumped dye-laser. Two nitrogen laser pumped dye-lasers werc in operation. They are in the process of attempting line narrowing by etalon and power amplification using a second dye cell pumped by the same nitrogen laser. The peak power of their small nitrogen laser is reported to be ≈ 500 kw.

They are doing quite well in getting the lasers to run. But again the situation is similar to other institutions—they have not yet carried out extensive laser chemistry research. The collaboration of the faculty members of the Chemistry Department and Optics Department will certainly be fruitful. In the Optics Department, there are two additional research groups we visited: the group on light sources and the group on vacuum electronics. These two groups are doing extremely well and the level of technology is very high. If the Chemistry Department's faculty members obtain their technical assistance and carry out some collaborative research, a very sophisticated chemical physics program could develop here.

There are many other institutions that are interested in setting up laser chemistry groups. Especially worth noting are the Institute of Chemistry in Peking and the Institute of Chemical Physics in Talien. These are the leading institutions that will engage in microscopic chemical kinetics in the near future. The University of Science and Technology operated by the Chinese Academy of Sciences in Hofei (in Anhui Province) is another institution that may play an important role in laser chemistry. All three institutions are also planning to initiate molecular beam research. The scientists in these institutions, although they have not actually worked on laser chemistry research, seem to be very well informed on the current literature.

Another area in which they have done relatively well is the fabrication of various crystals for laser operation. In the Institute of Physics in Peking and in the Institute of Ceramic Chemistry and Technology in Shanghai, we saw various crystals for laser generation and for frequency doubling such as ruby, Nd-YAG and lithium niobate crystals. In the Institute of Ceramic Chemistry and Technology a Nd-YAG system with a doubler was in operation for checking the quality of the crystals that had been grown at the Institute.

The area of laser chemistry, like many other areas of basic research, is rather weak at present. Chinese determination to develop the laser industry and actively engage in research in this area will undoubtedly bring some rapid progress. The lack of a large number of experienced, leading scientists probably will be the initial obstacle to rapid development in the future.

E. Isotope Separation

Professor Chang Ch'ing-lien, Chairman of the Chemistry Department of Peking University, and Wang Te-hsi, Deputy Director of the Institute of Atomic Energy, Peking, have been the leaders in isotope chemistry in China. Chang, well known in the literature of isotope chemistry, started a program to enrich deuterium and ^{18}O in water in 1956 at the Academy of Sciences based on his work at the Nobel Institute in Stockholm in 1938. Oxygen-18 is still enriched by water distillation somewhere in the northwest of China. In 1960 Chang built a Spindel-Taylor nitrox plant at Peking University. That plant was subsequently transferred to the Institute for Chemical Engineering under the Ministry of Chemical Industry in Shanghai. It is producing 99.5% ^{15}N on a laboratory scale. In 1960 Chang started work on separation of ^{10}B by the BF_3-anisole process. He found extensive decomposition of the anisole and switched over to the BF_3-diethyl ether complex (British process). There is now a large-scale boron isotope separation plant in Talien. Chang is interested in setting up an NO distillation plant for enrichment of ^{15}N and ^{18}O. He has available a Varian Mat CH5 isotope ratio mass spectrometer and would like to exchange standard water samples for intercomparison of absolute standards on deuterium and ^{18}O. We suggested that perhaps arrangements could be made for him to get some samples from the National Bureau of Standards.

The isotope separation processes that have been placed into production are based on fundamental work carried out in the United States prior to 1960 and are derivatives of U.S. technology of the 50s. Chang, Wang, and Wu, Director of the Kirin Institute of Applied Chemistry, are realistic about their present science and technology and plan to advance this field in the new plans for Chinese science and technology. Toward this end there will be university graduate programs in isotope chemistry at Peking University and isotope separation technology at the Technical Institute at Tientsin, as well as a program in isotope geology involving cooperation between the Glaciology Institute of Lanchow and the isotope chemistry program at Peking University. In addition Professor Li Kuan-hua will initiate a graduate program at Inner Mongolia Teachers' College in areas relevant to paleothermometry, an interest that he has pursued actively since he received his Ph.D. at the University of Chicago in 1950 with H. C. Urey for work in this area.

There is intense interest in China in the potential use of crown ether complexes in liquid–liquid chemical counter current exchange processes. The Chinese were impressed with the separation factor of 1.0010 reported by Jepson and Dewitt (*J. Inorg. Chem.* **1976** *38*, 1175) for $^{40}Ca/^{44}Ca$ separation. They plan to pursue the potential of these methods for the separation of the isotopes of the light elements at the Institute of Atomic Energy, Peking, and are intensely interested in getting details about the South African and the new French chemical exchange processes (Coates, San Francisco Conference, December 1977). With respect to the latter, they were interested in a reconciliation of the reported separation factor $\alpha - 1 \simeq 3 \times 10^{-3}$ with typical values of $\ln(s/s')f$ for uranium components of 2×10^{-3}.

Electromagnetic Separators

There are three electromagnetic separators at the Institute of Atomic Energy outside of Peking. Two of them were built in the Soviet Union and placed into operation in 1965 and 1970. The third separator was built in China. The separators that were delivered from the Soviet Union had poor ion beam focusing. The Chinese shimmed the magnets and increased the resolution by a factor of about 3. Typical ion source currents are of the order of 200 to 300 mA. In general the source operates at about 1000°C. Mr. Chang Wei-ming showed us a collection of some of the samples separated in the electromagnetic separation program. These were ^{25}Mg, ^{26}Mg, ^{42}Ca, ^{44}Ca, ^{58}Ni, ^{62}Ni, ^{63}Cu, ^{65}Cu, ^{112}Sn, ^{116}Sn, ^{50}Cr, ^{53}Cr, ^{84}Sr, ^{87}Sr, ^{168}Yb, ^{171}Yb. Ytterbium is the heaviest element for which they have separated isotopes. The purity of ^{168}Yb and ^{171}Yb are 12.5% and 92.8%, respectively. They also showed us a sample of a few mg of ^{48}Ca that had been enriched to 30%–40% isotopic purity.

Laser Isotope Separation

There is strong interest in laser isotope separation in China. Programs are being initiated for ^{13}C enrichment by multiphoton absorption at the Kirin Institute of Applied Chemistry (Ch'angch'un), and ^{10}B separation at the Institute of Physics (Peking). There is also work at the Institute of Optics in Shanghai. Scientists at the Institute of Saline Lakes, Hsining (Tsinghai Province) are reported to have been working on Li isotope separation by lasers for two years.

Uranium Enrichment

There is reported to be a full-scale plant to enrich natural uranium to weapons grade ^{235}U in the vicinity of Lanchow by the gaseous diffusion process. Like all gaseous diffusion plants (U.S., France, and U.S.S.R.) the Chinese plant is a "classified" area and was not part of our program. The Lanchow area is a reasonable area to site such a facility. It has been developed to a highly industrialized area; it has ample cheap electric power from coal generating plants and hydroelectric plants on the Yellow River, and the area lends itself to isolation.

F. Instrumentation

It is evident in high-priority areas, particularly the applied research areas closely related to major national priorities, that research institutions in chemistry and chemical engineering are able to obtain the facilities required to support their missions. We assume that must also be the case in sectors we did not see, such as defense and space.

Most of the simpler laboratory instruments available appear to be of indigenous design and manufacture. For example, gas and liquid chromatographs, electrochemical instrumentation, high-vacuum hardware, pumps, flanges, recording instruments, meters in refinery control rooms, etc., all are of Chinese manufacture. Machined parts for high performance requirements using unusual materials are well made and are comparable in quality to similar components in the United States. A variety of electronic test instrumentation of Chinese manufacture such as scopes and meters of all types is also available. Electronic instruments carry labels indicating manufacture, for example, in Shanghai, Kirin, Ch'angch'un, Harbin, Lanchow, Anshan, etc., indicating a widely distributed and diversified base for instrumentation manufacture. Most complex instrumentation such as NMR, mass spectroscopy, automated X-ray diffraction, and electron microscopy still appears to be imported, with Japan having the largest share of the market.

What appears to be missing in the instrumental area are the most sophisticated instruments with integrated microprocessing available in the United States, Western Europe, and Japan in the past five years. This category includes, for example, Fourier transform NMR spectroscopy. There is apparently only one such instrument in China, a Varian XL-100 at the Institute of Photography, Peking, which is also the only instrument in China with ^{13}C capability. There is a high-resolution mass spectrometer located in the Institute of Chemistry, Peking; a laser Raman located at the Petrochemical Research Institute, Peking; and an automated X-ray diffractometer at the Institute of Biophysics in Peking, the only four-circle instrument available in China. In addition, we did not see any instrumentation for microwave spectroscopy, Mössbauer spectroscopy, ESR,* ESCA,* LEED, Auger, or EXAFS spectroscopy. Currently there are no programs in molecular beam research, although the Chinese are interested in initiating a molecular beam program at the Institute of Chemistry, Peking, and at the Institute of Chemical Physics, Talien. We did not see instrumentation for circular dichroism or automated scintillation counting. Most instrumentation (with a few exceptions) operates without the benefit of computer interfaces.

Institutions that appear to have adequate instrumentation to support their missions have high-priority programs with applied emphasis, which were generally spared the turmoil of the Cultural Revolution. These include, for example, the Institute of Ceramic Chemistry and Technology, and the Catalysis Research Institute of Futan University. At the Peking Petrochemical Research Institute the instrumentation is adequate to support the mission of the Institute, with emphasis on control and development functions. The work of this Institute is supported by three pilot plants including a remarkable pilot fluid-bed cracking unit about 40 m in height. What appeared to be missing in these pilot plants are instruments widely used in the United States for on-line analysis and control. With this exception the instrumental facilities in this institute are roughly the equivalent of what would be found in a development laboratory in the United States associated with a small chemical or petroleum company.

Petrochemical research is clearly an area of high national priority. This is evident not only from the facilities available and the general housekeeping at the Petrochemical Research Institute, but also from the good quality scientific manpower available and the stable staff and working conditions even through the turmoil of the Cultural Revolution. The Director of the Peking Petrochemical Research Institute apparently reports to the Manager of the Peking Petrochemical Works, who in turn is responsible to the Ministry of the Chemical Industry.

* There are reported to be home-built ESR spectrometers at the Institute of Chemistry, Peking and at the Institute of Chemical Physics, Talien, and an imported ESCA spectrometer at the Institute of Organic Chemistry, Shanghai.

Instrumentation at the Peking Atomic Energy Institute includes, for example, Chinese-made oscilloscopes, multichannel analyzers imported from France, spectrometers imported from England and East Germany, and an AEI mass spectrometer interfaced with a PDP8(E) computer (sold through the English subsidiary of Digital Equipment Corporation). Solid-state detectors for scintillation counters are available, but NaI detectors are in the most frequent supply. The Atomic Energy Institute appears again to have enjoyed high priority and stability.

At the Institute of Nuclear Research in Shanghai, a multichannel analyzer constructed with integrated (MSI) circuit boards was observed. The design and layout as shown in Fig. 11 appear to be quite good. Again this indicates that in high-priority areas the Chinese can and do make available technology that is at the forefront.

One interesting activity at Kirin University in Ch'angch'un is the semiconductor laboratory in the Department of Semiconductors. This is a laboratory that was established to train students in the design and production of medium-scale integrated circuits. It includes a variety of equipment beginning with design and layout of circuits, production of mats, reducing these and applying images to photoresistive materials, processing under clean environmental conditions, and the final assembly of completed chips. The facility looked well run; we are not aware of a comparable approach in the United States for the training of third-year undergraduate students in such techniques. This appears to be a reasonable approach for creating a body of trained manpower for the development of a large-scale integrated circuit industry.

The Institute of Materia Medica in Peking is one of the institutes that appears to have limited instrumentation to support its mission. For example, the NMR work at this institute is being carried out with a 60-MHz JEOL instrument. Although the application of this instrument is for proton spectroscopy only, the use is fairly sophisticated, including use of the Overhauser effect and chemical shift reagents in chemical structure determination. Some of the other structural methods that one might expect to find in a natural products laboratory are available, but major sophisticated instruments for NMR and mass spectroscopy that would be found in a comparable laboratory in the United States are missing. The Institute of Materia Medica is managed by the Chinese Academy of Medical Sciences, which is in turn under the Ministry of Public Health. The development of pharmaceuticals from natural products is again an area that appears to have enjoyed priority and relative immunity from political turmoil.

In institutes that have a basic research mission, and particularly in university laboratories, the instrumental facilities available are limited or inadequate by U.S. standards. There are striking differences, for example, between facilities available at the research institutes described above and those available at Peking University and Tsinghua University. These two universities were hard hit by the turmoil of the Cultural Revolution and basically were unable to carry out research or teaching programs for the 10-year period from 1966 through 1976. Facilities for research are very limited at these two universities. However, the university faculties are well aware of this problem and are anxious to rebuild their facilities. Those institutions whose focus is in basic research evidently have been less favored than those with a clear applied research and development mission.

The shops and support facilities in the various institutes visited appear to be of excellent quality. For example, the Institute of Biophysics, Peking, operates a machine shop (shown in Fig. 20) and a glass shop that not only support the research operations at this institute but also provide services for other institutes as well. The machine tools, including lathes, milling machines, grinding machines, etc., appear to be of high quality and the machinists seem to be well qualified. The glass shop included four glassblowers, all of whom appeared to be competent, and the finished work was of good quality. Stainless steel welding also appeared to be of good quality. The shops appeared to be more extensive than would be required in the United States to support the research programs that appeared to be underway at these institutes. The shops presumably assume a general maintenance function that is more significant than comparable shops in the United States due to the lack of manufacturers' representatives to help carry out maintenance functions. The shops also have a major role in

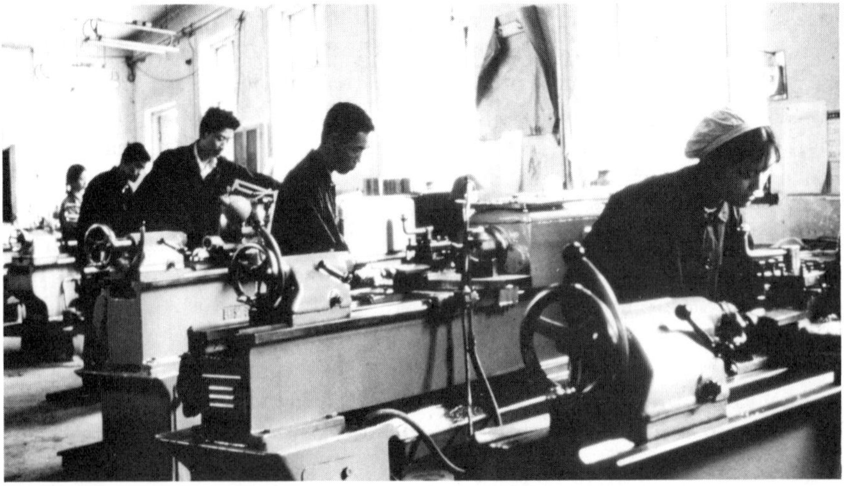

Figure 20. Machine shop of the Institute of Biophysics in Peking.

the manufacture of spare parts (which normally would be available directly from manufacturers in the United States) and may also provide specialized services to local industry.

The development of instrumentation of minimum sophistication appears to depend on the initiative of individual scientists in the various research institutes working in collaboration with instrumentation factories. Instruments developed in this way include gas chromatography, high-pressure liquid chromatography, polarography, vapor-phase osmometry, and instrumentation for the measurement of boiling point elevation. Typical instrumentation factories include the Analytical Instrumentation Plant (Tientsen), the Analytical Instrumentation Plant (Shanghai), the Peking Analytical Instrumentation Factory, etc. Prototype instrumentation of this type is generally first constructed in the extensive shops associated with each research institute. Prototype instruments are then produced in small quantities by the instrumentation factory and distributed to various research institutes for testing and improvement. Normally the research institute shops and the analytical instrumentation plants are responsible for making almost all of the components—pumps, seals, etc.—in addition to assembling the final instrument. During the period of the Gang of Four, the development of instrumentation was evidently favorably regarded, providing substantial incentive for instrument development as opposed to basic research.

Instrumentation of greater sophistication such as scanning electron microscopy (Fig. 17), gas chromatography/mass spectroscopy, and infrared spectroscopy, has evidently been developed in an Institute for Scientific Instrumentation associated with the Chinese Academy of Sciences. This group includes electronic and mechanical design capability for larger systems, as well as facilities for prototype manufacturing. We observed several instruments produced in this way, including a GC/MS instrument under prototype testing at the Institute of Chemical Physics in Lanchow shown in Fig. 21. One particularly ambitious project is an attempt to build a Fourier transform NMR spectrometer based on the Varian XL-100 design at the Kirin Institute of Applied Chemistry. The magnet and console for this instrument are shown in Fig. 22.

There appeared to be very little instrumentation of original design that might be of interest to U.S. laboratories. However, one example is a microwave device used to measure the rate of diffusion of water through oil-well core samples. In this device microwave radiation at 25 GHz from a klystron source is passed through attenuators and level measuring components and through a core sample that is held between a pair of transmitting and receiving microwave horns. Since water in the sample produces a high dielectric loss, the measurement of the intensity of transmitted microwave radiation

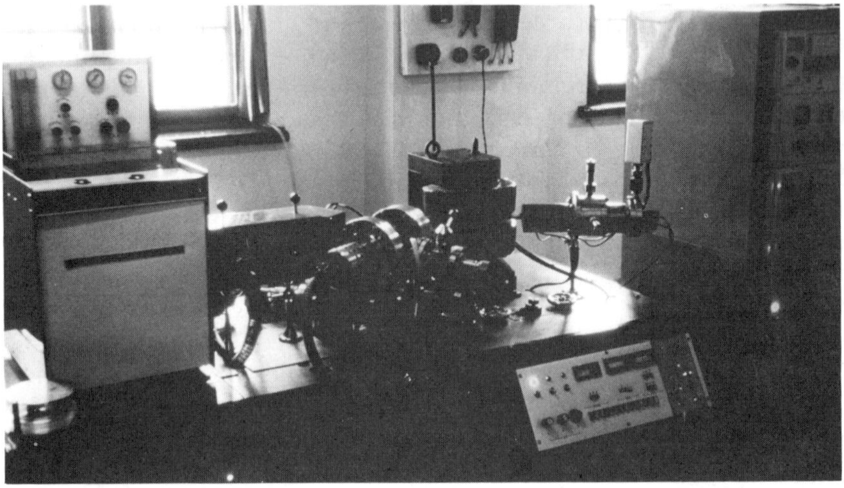

Figure 21. GC/MS under prototype testing at the Institute of Chemical Physics, Lanchow.

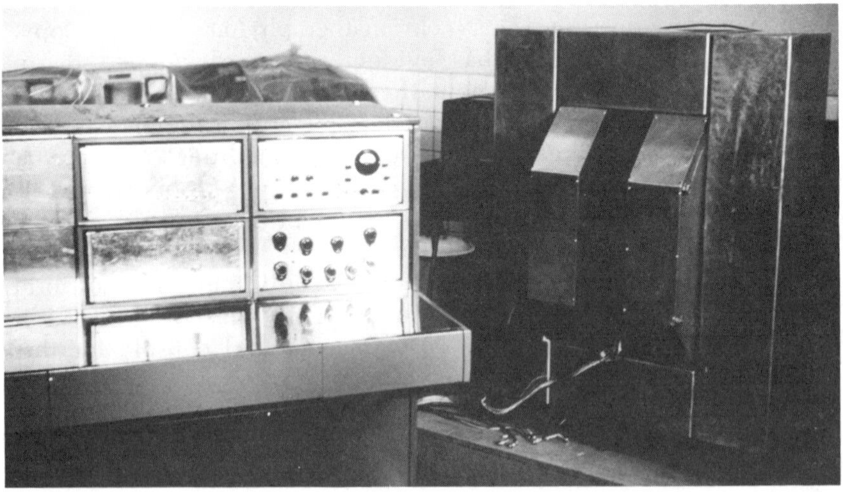

Figure 22. High-resolution NMR spectrometer under construction at the Kirin Institute of Applied Chemistry.

provides a simple measure of the quantity of water diffusing through the core sample. This microwave device shown in Fig. 23 was produced at the Tach'ing Academy of Science, Technology and Design Laboratory of Fluid Mechanics.

Another interesting device is a multifunction instrument for testing water samples developed at the Kirin Institute of Applied Chemistry. This instrument was designed to measure the water temperature, pH, conductivity, and concentration of specific ions including cyanide. The device was compactly packaged with various probes and a selection of pushbuttons to register each of the above variables.

Of particular note is the construction of a cw argon-ion laser coupled with a jet stream dye laser at the Kirin Institute of Applied Chemistry. This laser is coupled with a double grating monochromator (Zeiss, Jena, manufactured in East Germany). This laser system represents a significant achievement. Other evidence of the successful efforts in the laser field include a high power pulsed CO_2 laser constructed at the Kirin Institute of Applied Chemistry and the pulsed CO_2 laser in use at the Institute of Physics, Peking, for isotope separation studies.

Figure 23. Apparatus for measuring water diffusion rates in oil well core samples by microwave dielectric loss at the Tach'ing Academy of Science, Technology and Design Laboratory of Fluid Mechanics.

Other examples of potentially interesting instrument developments include the gas chromatography work at the Institute of Chemical Physics, Talien, which has produced a flame ionization-electron capture detector of new design and good quality. Also C–H analyses were being carried out in the Institute of Organic Chemistry in Shanghai using an instrumental method based on a coulombic measurement.

Instrumentation for teaching purposes is generally up to the standards of the U.S. undergraduate teaching programs in Futan university in Shanghai and was adequate at Kirin University in Ch'angch'un. Instrumentation for undergraduate laboratory courses was very limited at Peking and Tsinghua Universities in Peking.

Top-quality scientific instrumentation is extremely scarce in China and thus good cooperation is essential. However, it was our general observation that no reasonable system exists for assuring access for Chinese chemists to the instruments they need. People in institutes that had a particular advanced piece of equipment generally indicated that their own work took first priority, and after that personal connections often played a role in determining who else would have access to the instrument. This is one place where the highly organized Chinese society could use a little more organization, namely user committees to assign priorities for use of valuable instruments that are in short supply and that are needed by chemists all across the country.

In order to support the increased level of effort in basic research projected in the 8- and 23-year plans, as indicated by the four modernizations and the eight priority areas for science and technology, the Chinese will need a substantial infusion of foreign instrument technology, at least in the short term. They can be expected to develop further the infrastructure to maintain these instruments and to produce their own instruments based on these foreign designs. The instrumentation area appears to be a high-priority area for expenditure of hard currency reserves. A substantial and diversified infrastructure appears to be in existence for design and manufacture of instruments of moderate sophistication. Even in the short term the Chinese have the capability to produce the instrumentation they need in high-priority areas.

G. Computers

The Chinese clearly have available a number of different types of computers, and they are able to use them in a variety of ways. However, there appear to be few computer systems in China that are capable of handling large data bases. Thus, there seems to be limited capability for the Chinese to use the computer tapes from *Chemical Abstracts* or the Cambridge crystallographic tapes or other large data bases of interest to chemists. Awareness of these resources will increase as the Chinese send young people for advanced training to major centers throughout the world. In the interim, progress will be slow in those areas of chemistry that are computer-dependent, such as protein crystallography, chemical crystallography, analysis of complex kinetics, and *ab initio* quantum mechanics. For the most part, Chinese chemists at this juncture do not have easy access to big computers and appear to be generally unaware that programs in the West could be available to them, for example, programs for computer simulation of NMR spectra or crystallographic computing packages.

The minicomputer appears to be an important new development in China. For example, the GC/MS unit at the Institute of Chemical Physics, Lanchow, comes with a computer having a 4K memory with 16-bit words, and a magnetic drum of 48 kW. However, we did not observe any examples of computers being used in process control, although a number of places are said to have programs of research in automatic process control.

Some of the computing facilities available at various institutes are described below:

Institute of Chemistry, Peking

It appears that there are no general purpose computer facilities in the institute of Chemistry itself. Rather nearby on the same site there is a computer called the "TQ-16," which is a 32 K machine with a 10-microsecond cycle time. This computer reads paper tape but not magnetic tape. It has an ALGOL 60 compiler but no FORTRAN compiler. There was reference to a Model 013 computer with a 0.5-microsecond cycletime available at the Computer Sciences Institute, but apparently few people at the Institute of Chemistry have availed themselves of this computer. No information was obtained on how access to these various computers is determined. A PDP-8E plus disc attached to the AEI spectrometer in the Institute of Chemistry is potentially more powerful than the TQ-16 now being used, but it does not appear to be used for general purpose calculations.

Institute of Biophysics, Peking

Apparently the Institute of Biophysics has no computer facilities of its own but makes use of the nearby facilities, which consist of a computer with the designation 109-C013 that has a 110,000 word memory and a 10-microsecond cycle time. In further discussions it was reported that the 013 now has a cycle time of 0.5 microseconds, which would be consistent with what we learned at the Institute of Chemistry. For the calculation of structure factors on insulin—8000 structure factors, 800 atoms—it took about one-half hour. That would place the machine probably in the 6400 class of CDC, but we could not be sure this applies to the machine as it existed a few years ago at 10 microseconds or whether it is the 0.5 microsecond time instrument. We did not obtain any information on the presence or absence of parallel processors in the present machine.

Talien Institute of Chemical Physics

The computer center at the Institute of Chemical Physics has a DTS-130 computer made in China. It uses 16-bit words and operates on BASIC. The core memory is 16 K. The cycle time is of the order of 10 microseconds. Input is by means of punched paper tape. They do have a line printer for output and also a simultaneous character display on an oscilloscope screen. They have done some X-ray structure calculations using this computer and also work on the optimization of linear circuits. A new and larger computer is being planned for this institute, a system manufactured in Shanghai with a cycle time of 1 microsecond. They are also initiating computer network studies and automatic information retrieval, and planning has been done for remote access terminals in the future. The time for planning of computer-to-computer connection is somewhat further in the future.

Futan University

The present computer, a model 719, Shuzi Dianzi Jisuanji, was installed in September 1971. This computer has a 125,000 word storage, each word containing 48 bits, and it has a cycle time of 2.4

microseconds. It depends entirely for input on paper-tape readers, although it does have a nonconventional magnetic tape for intermediate storage and for secondary input. This tape appears to be 16-track. The machine itself is solid state, and output consists of line printers. There are no graphics, no cards in or out, no remote facilities, and the compiler is ALGOL; there is no FORTRAN compiler. The machine is saturated and runs full time. By 1980 the University plans to have a larger computer that will include remote facilities, faster computations, and graphics. It was not apparent that anyone in the Chemistry Department was making heavy use of the computer.

Shanghai Institute of Chemical Engineering

The Shanghai Institute of Chemical Engineering has a computer, a Chinese model CT-709, which is made in Shanghai with a cycle time of 8.0 microseconds using a 48-bit word floating point. It has a 32 k-word memory, with four external magnetic drums at 48 k-words per drum, a paper tape input but no magnetic tape, and a printer and X-Y plotter. It uses ALGOL language with a modified FORTRAN converter. It has no remote terminal.

Shanghai Science and Technology Computing Center

Currently the Shanghai Science and Technology Computing Center uses the 709 computer (32 kw of storage, 48-bit words, and 8.0 microsecond cycle time) and the 731 computer (64 kw, 48-bit words, 5.0 microsecond cycle time, and the 64 kw core is divided up into blocks of 32 kw). The 709 computer is essentially the same as the TQ-16 manufactured in Shanghai. The 731 has roughly the following configuration: 5 tape units, some of which are of French design and are 16-track; 2 drums, both used for intermediate storage; 5 line printers and 2 paper-tape readers and writers; a wide bed XY recorder, which is run directly off the computer. The power in Shanghai is relatively stable and the computer generally undergoes maintenance for 2 hrs per day.

It is clear that this institute has been successful, that their computer capability is saturated, and that they are looking for a larger computer. There is active cooperation with various chemical institutes including, for example, collaboration with the Institute of Biochemistry on the mechanism of enzyme reactions. It also appears that chemists who are aware of this institute do come to seek help. The problems now being solved seem to be of increasing complexity.

The institute charges its customers a small fee primarily for computer maintenance. There is no charge for personnel services. The TQ-16 or 709 is described as a medium-sized computer. The 109-C013 computer is considered to be one of the largest computers in China. The Chinese indicated interest in data bases but noted that the memory on their present computer was too small to allow them to consider it. They appeared to be unaware that they might obtain programs from overseas. They said they would like a better computer that is able to do at least 2×10^6 calculations per second, to be stable, and to have more core and convenient input–output devices. They indicated that convenient input–output devices would include a cathode ray tube output (a graphics package), and a more convenient input, including magnetic tape and magnetic discs. They did not appear to be knowledgeable in remote access techniques but indicated a strong interest in more sophisticated compilers, including FORTRAN and COBOL. The fact they want COBOL indicates an interest in data processing and in business operations.

The Shanghai Science and Technology Computing Center has concentrated on a popularization of computer applications. For example, they will go out to factories or institutes to encourage people to use programs they have written that have particular applications, they will teach people how to use a computer and how to program it in case they need to write programs of their own, and they will also assist in numerical analysis.

7. ORGANIZATION OF CHEMICAL RESEARCH AND DEVELOPMENT IN BROAD NATIONAL MISSION AREAS

A. Energy

Introduction

"Energy resources" ranks third on the list of national priorities of China recently pronounced by Fang Yi, the Director of the Commission on Science and Technology. This indicates that the Chinese are well aware of the importance of energy as a component of their long-term development strategy. At present there is no energy ministry analogous to the Department of Energy in the United States; however, there are active programs in the various ministries with responsibility for energy-related programs. These programs are coordinated through the State Planning Commission and through the No. 2 Bureau of the Commission on Science and Technology. Programs are underway in several key energy-related areas, and these appear to fit well into a broad national plan. Descriptions of facilities and programs observed in various energy-related areas are provided in the following sections.

Petroleum Resources

In 1959 a major Chinese oil find occurred at Tach'ing, the northeast section of China approximately 240 kilometers northwest of Harbin. A typical producing well in this field is shown in Fig. 24. Prior to this strike, China had only limited internal oil supplies (Yumen fields) and shale and coal were major sources of energy and chemicals. The availability of an indigenous petroleum supply has generated the impetus for energy self-sufficiency in China and laid the framework for a rapidly developing petrochemical industry.

Figure 24. Producing well in the Tach'ing Oil Field.

Supply. Accurate Chinese oil production figures are not available. Tach'ing is, however, the major producing field in China today and Table II estimates Tach'ing oil flows to various petroleum refineries and chemical plants located from Shanghai to Tach'ing. The total flow aggregates some 27.2 million tons/year, excluding oil exported. On a recent trip to China Secretary Schlesinger of the U.S. Department of Energy reported that the Tach'ing oil field production rates total 50 million tons/year.

The Tach'ing oil fields are being depleted at a rate of 2.5% per year, yielding 40-year field life. Since the field produces at a rate of 27.2 million tons/year to 50 million tons/year, the Tach'ing reserves

TABLE II
Tach'ing Crude Destination

Location		Crude Flow, Tons/Year (millions)
Tach'ing		4
Talien		5
Peking		7.5
Shanghai:	Chemical refinery	1.7
	Petroleum refinery	5.0
Fushun		4.0

can be estimated at (27.2 to 50) × 40 or about 1.2 to 2 billion tons (9.0 to 15 billion barrels at 7.5 barrels/ton). This reserve figure may be compared with the figure of about 10 billion barrels of the U.S. North Slope. Approximately 25% of these reserves have been produced but further exploration at Tach'ing is still underway. The field is said to be producing at a rate some 50 times its initial 1959 rate. This field is only one of several that are stated to be large. Others include one in Sinkiang Province, one at Takung near Talien, the Po Hai offshore fields, and Shantung (Shengli). China has as an objective the discovery by the year 2000 of 10 oil fields the size of Tach'ing. The Chinese are extremely proud of their achievement in oil exploration and production, and the Tach'ing oil field experience is expressed in their motto concerning industrialization, "Learn from Tach'ing."

Production. The only oil field observed was Tach'ing, and this oil field consists of "several thousand" naturally flowing wells averaging 20–100 tons/day of production with an average depth of about 3000 ft. The production wells have water reinjection to stabilize the field pressure and the field has a remarkably loose formation said to be four darcies. At Tach'ing the well spacings average two per acre. The Tach'ing crude has a high paraffin content and must be heated to prevent wax precipitation. Heating is accomplished by separation and burning some of the associated dry gas. This gas averages about 3200 m^3/well/day. The Tach'ing crude is low in sulfur ($\approx$ 0.1%) but contains arsenic, which presents a refining problem. It is the only Chinese crude with an appreciable arsenic content. The arsenic is apparently removed in a Co/Mo hydrotreating step and buried with the spent Co/Mo catalyst.

Transportation. Transportation of Tach'ing crude requires heated pipelines and tankers. A transportation infrastructure is being built and there are, for example, a 1600-km heated pipeline carrying Tach'ing crude to Peking, a second pipeline to Fushun and to the new port facility in Talien, and heated insulated tankers carrying Tach'ing crude to the refineries at Shanghai. Other crude is not so paraffinic. For example, at least 2 million tons/year of Sinkiang crude travels 2000 km by unheated railroad tank cars to the refinery and chemical complex at Lanchow. The railroad tank cars have a 50-ton capacity requiring the movement of some 40,000 tank cars of crude per year from Sinkiang. A pipeline is said to be planned, but construction has not started.

The Chinese have also constructed facilities for the export of crude oil at Talien. These facilities began operations in 1976 with

Figure 25. Port facilities in Talien for loading tankers.

equipment designed and built in China. The port consists of a 300,000-ton storage depot, a dock some 1.7 km long capable of loading Chinese-built tankers shown in Fig. 25 (made in Canton) of some 100,000-ton capacity at a loading rate of 10,000 tons/hour. They currently export Tach'ing oil from this port to Japan and Southeast Asia, and they are interested in exporting oil to the United States.

Hydrocarbon Conversion

With the recent intensification of the thrust to develop its petroleum resources, China has been constructing a relatively modern petroleum processing capability. Their activities can be broadly divided into two main processing areas: petroleum refining for fuels and lubricants, and petrochemicals. It is helpful to compare the Chinese activities in this regard to a modern, fully integrated petroleum refinery producing fuels, lubricants, and chemicals. The flowchart for a modern U.S. refinery is shown in Fig. 26.

From the following discussion it will become obvious that the Chinese have most of the elements of such a modern refinery. In

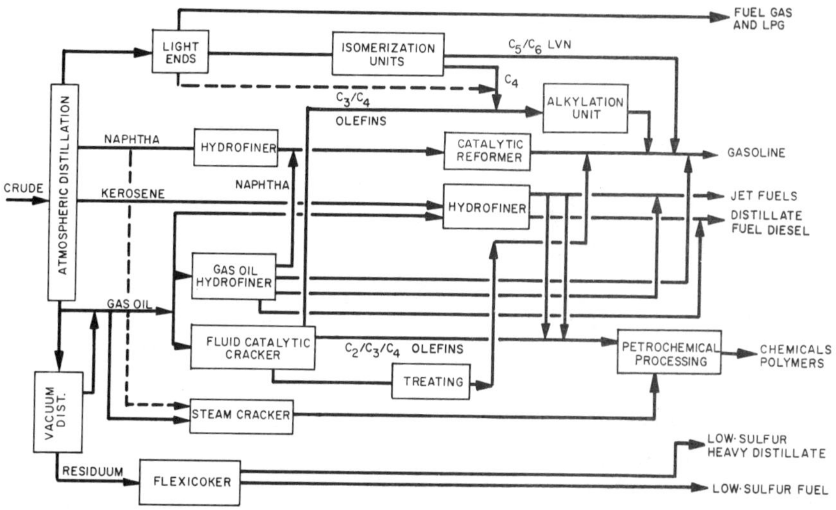

Figure 26. Flowchart for a modern refinery.

some categories, their technology is somewhat out-of-date, but in most cases their technology is quite modern. The most obvious gap involves the absence of computer control within the refinery/petrochemical complexes.

Refineries visited are shown in Table III. Except for the Fushun refinery, which was originally a plant for producing synthetic oil from shale, the refineries are less than 20 years old (including the Shanghai Chemical Refinery). Most of these refineries have a chemical complex associated with the refinery. The plants were designed and constructed by the Chinese using modern techniques (Fig. 27) with only limited outside aid. As far as practicable Chinese-built machinery is also used throughout the refineries, for example, pumps (Fig. 28), heat exchangers (Fig. 29), and cooling towers (Fig. 30). There is one exception

TABLE III
Approximate Year Construction Initiated at Refinery Visited by Delegation

Refinery	Year
Tach'ing	1959
Peking	1964
Talien Refinery Plant No. 7	1964
Fushun Refinery No. 2	1952 (originally shale)
Lanchow	1959
Shanghai (Chemical Refinery)	1965

Figure 27. Refinery construction technique employing guyed crane.

and this is the refinery at Lanchow. The Lanchow plant was built by the Soviets using Soviet designs and Soviet machinery. It is the oldest refinery of the modern era and has two TCC units with a combined capacity of about 500,000 ton/year. The other refineries have dense fluid bed catalytic cracking units of the type shown in Fig. 31 employing microspheroidal zeolite cracking catalysts. The units are designed and constructed in China and the catalyst is also developed by the Chinese, based on Western literature, and produced in China.

Such dense-bed reactors have been largely replaced in the United States and elsewhere by the riser cracker. The latter units are more efficient in their use of the unique properties of zeolitic catalysts. At the Peking Institute of Petroleum we observed a large riser cracker pilot plant and were told that the riser cracker has been developed and commercialized in China.

Most refineries had fixed-bed platinum reforming units of the type shown in Fig. 32 also designed and built in China. Since the Chinese refineries turn out a low-octane gasoline (70–83 ON), the main function of the reformer is a benzene, toluene, xylene (BTX) cut. The refinery at Lanchow had a multimetallic fixed bed reforming unit. Both the platinum and the multimetallic reforming catalysts were Chinese-developed. This is another instance of the Chinese desire to maintain an up-to-date refining infrastructure. We were told, for example, at the Peking Institute of Petroleum that they had performed the research work on multimetallic catalysts, which had led to semicommercial production. It is not known which multimetallic catalyst is in use in Lanchow, but the Peking group noted that their own system was a platinum/tin/alumina combination.

The refineries also had the usual atmospheric and vacuum distillation columns (Fig. 33) and some (Tach'ing/Fushun) had delayed cokers for bottoms conversion (Fig. 34). Propane deasphalting, solvent dewaxing, and extensive lube production facilities also exist

Figure 28. Refinery pump room at the Fushun No. 2 Refinery.

Figure 29. Heat exchangers at the Peking Petrochemical Complex.

Figure 30. Typical cooling tower at the Fushun No. 2 Refinery Complex.

at specific refineries. An interesting example of the ability of the Chinese to commercialize existing Western technology concerns a dilution chilling process for dewaxing at Lanchow. At Lanchow the technical people had read the Western literature on an Exxon process (Dill Chill) and tried it. They stated that "it worked," so they developed it for their own purposes.

Nowhere in China is HF alkylation or naphtha isomerization practiced. There is at least one sulfuric acid alkylation unit at the Lanchow refinery and a benzene alkylation unit at the same refinery that produces isopropylbenzene for use in aviation fuel.

Since China emphasizes diesel fuel, the Chinese refineries produce more diesel fuel than gasoline (about a 2:1 ratio). The automobiles have compression ratios of about 6:1 and require fuel with octane ratings in the 70–83 region.

Most refineries produce many products, as listed in the individual appendixes, besides gasoline and diesel fuel. Some (Lanchow, Peking, and Shanghai) appear to specialize in lubricants and lubricating oil additives, which are produced at these plants for distribution on a national scale.

The petrochemical industry in China is very young. Its genesis is coincident with the discovery of the Tach'ing oil fields in 1959. Prior to that time the production of chemicals was mainly coal-derived with some also produced from shale. An indication of the growth of

Figure 31. Fluid-bed catalytic cracking unit at the Peking Petrochemical Complex.

the industry is obtained by reference to Table IV, which delineates the petrochemical complexes visited and the year of their construction. The plants all have a less than 20-year history; two-thirds of them have been built within the past 10 years. There is very little Chinese-developed technology in evidence in the chemical complexes. This is a clear departure from their refining modus operandi and is undoubtedly a result of the rapid growth of modern chemical technology outside China. The Chinese attempt to manage jointly the construction of individual foreign-purchased process units and to use as much Chinese equipment as practicable in order to gain design, construction, and operating experience.

Figure 32. Fixed-bed platinum reforming unit at the Tach'ing Refinery.

Figure 33. Distillation columns at the Peking Petrochemical Complex.

Figure 34. Output from delayed coker at the Tach'ing Refinery.

TABLE IV
Years of Construction of Chinese Chemical Plants

Plant	Year
Tach'ing Petrochemical Complex	mid-sixties
Shanghai Petrochemical Complex	mid-sixties
Shanghai Chemical Plant	early seventies
Peking Petrochemical Complex	mid-sixties
Lanchow Chemical Plant	early sixties

Two out of four of the petrochemical complexes had a gas oil steam cracker. These units were partially designed by the Chinese but also used foreign technology especially in the heating coils.

Plant	Gas Oil Steam Cracking, t/yr
Tach'ing Petroleum Complex	Planning stage
Shanghai Chemical Plant	470,000
Peking Petrochemical Complex	>300,000
Lanchow Chemical Plant	?

Ethylene, propylene, butadiene, and aromatics were produced via steam cracking, separated and used for further chemical production. Catalytic reforming is used (except at the Shanghai Chemical Plant) to produce a BTX cut that is separated to yield benzene, toluene, and xylenes. The xylenes are separated using molecular sieve absorption to produce a *p*-xylene fraction, and at the Shanghai Plant at least, the *p*-xylene is oxidized to terephthalic acid, esterified with methanol to produce DMT, and reacted with ethylene glycol to produce polyester. At the Shanghai Chemical Plant the polyester is also spun into terylene fiber.

The Shanghai Chemical Plant is an interesting operation since it is a chemical refinery. Crude from Tach'ing arrives at the plant via heated tanker. The crude is distilled and a naphtha fraction sent to a nearby fertilizer plant where it is steam reformed and used as feed for ammonia production. The residual material is steam cracked and the products used to produce a range of intermediates that are ultimately used to produce fibers. For example, propylene is ammoxidized to manufacture the acrylonitrile intermediate; polyethylene is produced via high-pressure, low-density technology; polyvinyl alcohol is made via Wacker ($PdCl_2/CuCl_2$) oxidation of ethylene to acetaldehyde and terylene as described above. The products from the Shanghai Chemical Plant are summarized in Table V.

TABLE V
Shanghai Chemical Plant Products

Product	Capacity, t/yr
Polyacrylonitrile	47,000
Polyethylene	60,000
Polyvinyl alcohol	33,000
Terylene	25,000

All the other petrochemical plants observed were integrated into individual refineries, but all had modern technology (as noted in Appendix D) for petrochemical production.

The Chinese are experimenting with low-cost cotton substitutes. With a population of some 950,000,000, their needs for clothing are enormous, and they do not wish to take up land that could be used for food production to grow cotton. They have experimental low-cost polypropylene hose made from dyed polypropylene fibers (as well as polypropylene pencils)! However, they stated that vinylon (polyvinyl alcohol) and terylene (polyethylene terephthalate) will be the fibers of their choice in the future for production of a cotton substitute.

Styrene–butadiene rubber and polybutadiene appear to be the major rubbers. The butadiene feed is produced in part via gas oil steam cracking and in part by butene dehydrogenation. The polybutadiene is mixed with natural rubber and used mainly in tire production. A facility for drying polybutadiene is shown in Fig. 35.

There are apparently eight large ammonia synthesis facilities utilizing American technology (Fig. 36). In addition, a number of small ammonia synthesis plants are stated to be scattered throughout China. There are also large urea plants and the single urea plant observed at Tach'ing has a capacity of 480,000 tons/ year. Interestingly, the urea plant was operating at one-fourth capacity due to a shortage of railroad freight cars used to move products.

Figure 35. Polybutadiene drying unit at the Peking Petrochemical Complex.

Figure 36. Entrance to ammonia plant utilizing U.S. technology at Tach'ing.

In addition to plastics, polymers, and fertilizers, the Chinese produce a host of other chemical products. These include, for example, 2-ethylhexanol and other higher alcohols via hydroformylation for use in plasticizers, benzene both by toluene hydrodealkylation and by separating reformate BTX, ethylene glycol, and ethylene oxide.

In summary, the Chinese petrochemical industry, while young, has been built on the latest Western technology. Most plants observed were in the process of increasing both plant capacity and staff. There is every evidence that the Chinese have launched a modern petrochemical era and have every intention of using their indigenous petroleum resources to the fullest as chemical feedstock.

With regard to personnel, the Chinese petroleum refineries and chemical complexes are stretched to the limit. As an example, the Shanghai Chemical Works receives one out of every three engineers it requests and could use an additional 2000 engineers. Its current engineering staff is 100—a very small number for so large a plant. There are apparently three recognized schools turning out people with some petroleum chemistry, chemical engineering, or oil drilling skills: Tach'ing, Shantung (Shengli oil field), and Hupei. Both Tach'ing and Shantung provide all-inclusive training in oil exploration and production, refining, chemical machinery, and oil exploration and oil transportation. These are four- to five-year courses. The school in Hupei apparently concentrates on oil drilling. Other refinery chemical complex units have made arrangements to train people locally in petroleum engineering or chemistry.

Most of the refineries and chemical complexes comprise entire communities. The oil field and petrochemical complexes at Tach'ing, the Peking Petrochemical Complex, and the Shanghai Chemical Plant, for example, have been built with an intrinsic tie to the community. The plant manager appears responsible for hospitals, kindergarten and middle schools, apartment housing, and shops, although this is often a shared responsibility with the local province or community. It is, therefore, difficult to determine accurately how many workers actually run a given refinery. Some are engaged in operations, some in construction, some in transportation, and some in running the community. Thus, figures are meaningless unless broken down to specific tasks, which we did not attempt. However, the plants visited did not appear to be overmanned and looked, in terms of number of personnel in attendance, much like their U.S. counterparts.

Shale

There are two large-scale oil shale mining and refining efforts in China today: one at Fushun in the Province of Liaoning in northeast China and the other at Maoming close to Canton. The Fushun open pit mine, begun in 1914, is a source of both coal and shale. The first oil shale retort was built in Fushun in 1941 by the Japanese as the No. 1 refinery; subsequently a No. 2 refinery was built east of Fushun using a similar but improved technology.

Shale that contains more than 4.7% of organic material is called rich shale and below 4.7% is called poor shale. The rich shale is orange in color; the poor shale is gray. Only the rich shale is refined. The shale has 1360 kilocalories of combustion heat per kilogram, and consists of 75% ash, 14% volatile matter (including methane) 2.5% water, 3.7% fixed carbon, and 4.8% organic liquids. The oil shale retort is very similar in construction to a present-day Lurgi coal gasifier where 1- to 7-cm pieces of shale are charged from the top and air forced in from the bottom. The reactor shown in Fig. 37 has an inside diameter of about 3 meters and an outside diameter of about 4 meters. The space between is filled with heat-resistant bricks. The retort is 13 meters tall and the shale takes 9 hours to descend from the top to the bottom. At the bottom there is a grid that rotates at a speed

Figure 37. Oil shale retort at the Fushun No. 2 Refinery Complex.

of less than 1 revolution per hour. High temperature shale is discharged from the bottom into a pool of water and emerges into a slag pile. The retort operates at atmospheric pressure and the maximum temperature is 900°–1000°C at about 1 meter above the grid. No cooling water is used. At the No. 2 refinery there are 60 retorts divided into a front row and a back row of 30 each, and each row is further divided into units of 5 retorts. The whole plant makes a very impressive sight, but it has had no innovation from the original design.

The first oil shale unit was started in 1941 under the Japanese, and the second unit was built in 1954. The second unit processes 15,000 tons/day of oil shale rock with a production of about 120,000 tons of crude shale oil per year. It is said that the No. 1 plant at Fushun produces about 80,000 tons per year of oil, and the one close to Canton also produces about 80,000 tons per year. The product oil was formerly refined at the Fushun oil refinery, but since the discovery of the Tach'ing field, the Fushun refinery has been completely converted to refining Tach'ing crude and the oil shale products from the Fushun mine are turned over to small plants nearby. The spent shale contains about 2% coke that is left on the shale. It is believed that the exit temperature of the shale is 400°C. The Chinese seem to have no problem with the shale sticking together to make large balls.

It was reported that the slag is used for road fill, for mine fill, and for making cement. The waste water is treated to remove pyridine;

there is no good use for the rest of the by-products, which are discarded into a canal that eventually flows into the river. At Maoming, close to Canton, the same process is used but without coal nearby. The shale oil costs perhaps as much as 50% more to produce than Tach'ing crude, but the plant managers argue that the shale is coproduced with coal and that China must not waste resources. Since capital investments are depreciated over 50–100 years and there is no interest charged on fixed and working capital, the economics must be interpreted carefully.

A follow-up visit to these plants to discuss operating experience is recommended.

Coal

The only coal-related facility visited was the open pit coal mine close to Fushun. This huge open pit mine excavates coal and oil shale. Currently the production rate is 3.6 million tons per year of coal and 12 million tons of shale rock per year. The pit is 6.6 km long and 2.2 km wide with a current depth of mining of 270 m. The open pit mine was started in 1914. There is also a deep shaft mine in another part of Fushun as well as another open pit mine. There are in each of the mines tertiary and quaternary rocks that are 1 to 50 million years old. We were told by the Chief Engineer that there is a thickness of 30 m of soil, 300 m of a green rock layer, 110 m of an oil shale layer, and finally a coal layer of 120 m before igneous rock is reached. With these figures the final depth of the mine would be 560 m. These numbers do not seem to be consistent with a current depth of 270 m. Later on it was pointed out that these thickness figures were not measured vertically but as distances along the angle of incline, and the current angle of inclination is between 17° and 19°. A sketch of a cross-section of this mine as drawn by the Chief Engineer is shown in Fig. 38. A view of the mine is shown in Fig. 39.

During the war and in its aftermath, 800,000,000 m^3 of rock slag were left behind, damaging the mines. In 1948 the sides of the mine began to cave in. This was followed by water flooding, which further destroyed facilities. In 1958 Chairman Mao visited the Fushun Mine, giving renewed impetus to the work there. Present production was reported to be five times greater than before Liberation. The coal has a combustion heat of 7500 calories per kilogram, 5%–10% ash, 0.1% –0.6% sulfur.

Although China has large coal reserves, we did not see research programs related to either gasification or liquefaction of coal.

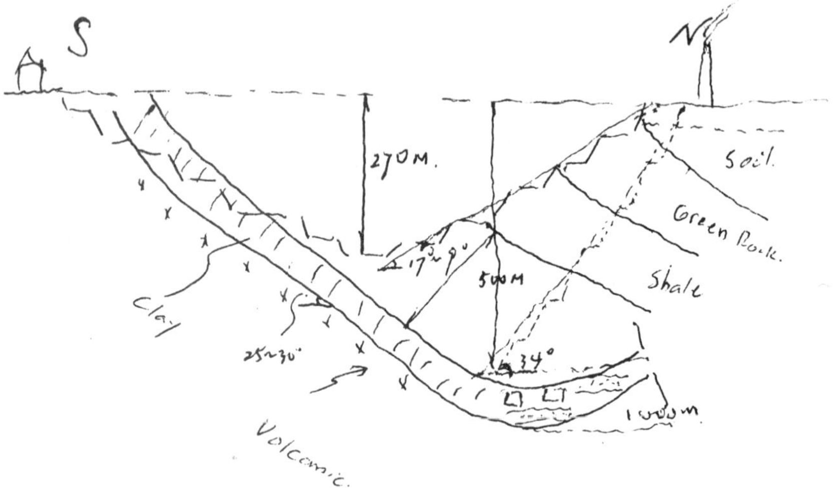

Figure 38. Cross-section of the Fushun open pit coal mine.

Figure 39. View of the Fushun open pit coal mine.

Nuclear Power

China has no nuclear power reactors at the present time. We were told that this source of energy has been delayed because it is related to their industrial base, which has not been adequate. They hope to learn from foreign countries in order to get started. However, the energy aspect of the National Plan includes emphasis on research in nuclear power generation and acceleration in the building of nuclear power plants.

The Institute of Atomic Energy, located near Peking, is responsible for investigating the various alternatives for nuclear power reactors and making recommendations to the central government. The research team there consists of about 40 engineers and reactor physicists. Planning for a Chinese national nuclear power program actually began at the Institute as early as 1974. They are apparently favoring the pressurized water reactor (PWR), but it is also possible that they may recommend some type of heavy water reactor.

The tentative plan seems to be that a prototype reactor will be built first, then a medium-sized power reactor, but the time schedules and sites have not been determined. The reactors would be Chinese-designed and constructed and would be operated by the Ministry of Water Conservation and Electric Power.

As an early step in the sequence, they plan to purchase one or two PWRs of the Westinghouse design in the conventional capacity range of 900 MW or higher, from France.

Development work on the reprocessing of nuclear power reactor fuel is also going on at the Institute of Atomic Energy. They are working to modify the Purex process used by the Chinese in their national plutonium production program for use in nuclear power fuel reprocessing. Besides research on modifying the chemistry of the process in order to effect improvements, they are investigating needed changes in the operating equipment. They are testing the use of centrifuges to replace the mixer–settler equipment that is used in their process for extracting plutonium at their plutonium production plant. A faster chemical separation process is needed in order to cope with the more radioactive nuclear power fuel resulting from the larger burn-up in nuclear power reactors.

Work is also underway at the Institute of Atomic Energy to find a satisfactory solution to the radioactive waste disposal problem. The view was expressed that solution to the waste disposal problem is the key to successful use of nuclear energy in China. (They find it difficult to understand why the United States, after so many years of study, cannot define a suitable plan to handle waste disposal.) There is an active research program to develop a process for the incorporation of medium-level waste in asphalt-like material. Consideration is being given to various methods of solidification of high-level waste, but research on this has not started as yet. Optimism for geological storage was expressed in view of the favorable data from the Oklo formation in Gaben.

Apparently the activities of the opponents of nuclear power in the United States are being watched closely. One of our hosts stated that the development and use of the breeder reactor in China is "inevitable."

Fusion

There is a small fusion energy development program in the Institute of Physics of the Academy of Sciences in Peking. The two facilities under development include a theta-pinch device (high-beta torus) that is under construction. At present there are large capacitor banks in place that are designed to produce 2.5 MJ, and the device will be designed to use a 16-kG field and to produce a plasma density of 10^{14}–10^{15}/cm^3. Expected temperatures are in the range of 100–600 eV, and containment times of the order of 100 nsec. Construction was started in 1977 and the current status is shown in Fig. 40. They are

Figure 40. Theta-pinch fusion device under construction at the Institute of Physics, Peking.

planning to use laser scattering and X-ray production as a measure of plasma temperature. A larger fusion containment device is apparently under development in Szechwan Province, and laser fusion technology is under development in Shanghai at the Institute of Optics and Applied Mechanics. The person in charge of this development at the Institute of Physics is Dr. Li Yin-an.

We also observed a working Tokamak in the Institute of Physics shown in Fig. 41. The ring appeared to be between 1½–2 m in diameter and the facility looked complete; it was either running or in a debugging mode. The toroidal field of the machine has a total energy of 3×10^6 joules and a 60-millisecond discharge time. The electron temperature of the plasma is about 3 million °C. The supporting instrumentation looked somewhat primitive by U.S. standards. The group operating the Tokamak indicated that there was a larger Tokamak under development at the "Southwest Institute of Physics" in Szechwan. The Szechwan Tokamak will be twice the size of the Peking instrument. At the present time no superconducting magnets are planned for the Tokamaks.

The scope of the program was clearly small and not appropriate for the development of a major program in fusion research. However, it did seem suitable for maintaining an awareness of developments in the fusion field and for providing a base of trained people with adequate experience to read the literature, attend international meetings, and recognize important developments abroad.

Figure 41. Tokamak at the Institute of Physics, Peking.

Solar Energy

In the inorganic chemistry division of the Kirin Institute of Applied Chemistry we were shown a multielement solar cell. The area of the cell is approximately 50 sq cm and it generates 450 mV. The cell is a composite of a base layer of cadmium sulfide that is vacuum evaporated onto a plastic mount and then coated with cuprous chloride (by being dipped into a solution of cuprous chloride). A protecting layer is put on next and the device is then gold plated. It is clearly not a prototype production model because of the high cost. The Chinese are presently studying stability of the cell. There was no evidence of large-scale silicon solar cell activity. The group at the Kirin Institute mentioned that solar cell work going on in other institutions included studies of silicon cells and also of other rare earth photovoltaic cells. There are ample quantities of rare earths in China, and there is research in many institutes looking for applications of rare earths. We were told that work on solar cells was going on at the Semiconductor Institute in Peking.

The persons we talked with regarded the production of economic solar electricity as a very long-range project.

Conclusions

Principal emphasis in the energy program in China appears to be placed on petroleum production and processing and coal production. The unique activity in shale oil production is being continued but not expanded. There is ample evidence of a well-thought-through program in the nuclear power area, and the Chinese appear to feel the time will come when they will require power from fission sources including breeders. The nuclear fission power program appears to be sensibly scaled for mid-term needs. The fusion and solar programs appear to be adequate to provide the Chinese a window into these areas of technology and are appropriate as a long-term component of a comprehensive energy program. All in all, the Chinese program in the energy area appears to be well conceived in terms of their short-, mid-, and long-term needs.

B. Materials

Materials science, as practiced in the United States, involves the study of metals, ceramics, and polymers, and complex systems involving combinations of these materials. In this context no interdisciplinary materials science institutes, laboratories, or educational programs were encountered.

Work in polymer science has already been described in detail. China produces most of the conventional polymers used as materials including polyethylene, polypropylene, polystyrene, ABS (acrylonitrile–butadiene, styrene), polyvinyl chloride, polybutadiene, polyisoprene, styrene–butadiene (SBR) rubbers, polyvinyl alcohol, polyacrylonitrile, polymethyl methacrylate, polycarbonate, polysulfone, polytetrafluorethylene, polychlorotrifluoroethylene, FEP and EPTE copolymers, silicones, polyethylene terephthalate, polyamides, and polyimides. No commercial production of polybutylene terephthalate or of polyphenylene oxide was apparent. While thermosets were available from Chinese sources, no university or institute programs for their study was encountered.

It was evident that there was technology for working with complex systems involving polymer-coated metal (teflon-coated frying pans, polymer-coated wires), although there did not seem to be extensive research in these areas in the laboratories visited. There also appeared to be available metal and fiber reenforced polymers (tires) as well as fiber glass-reenforced systems. There was institutional work on reenforced polyimides for high temperature applications.

"Materials Science and Technology" including the "updating of techniques for the production of plastics, synthetic rubber, and synthetic fibers" was one of the eight priority areas designated by Vice-Premier Fang Yi. With the development of the Tach'ing Oil Field, the growth of a petrochemical industry and the related growth of a polymer industry holds great promise. The technology of petroleum catalytic cracking and reforming appears to be well developed so that sources of a variety of monomers are available. There is strong emphasis on the study of polymerization catalysts and an emerging effort for the molecular and solid-state characterization of polymers. With the growing need in China for cheap materials for its large population and with the desire to reduce the dependence upon agriculture as a materials source, the expansion of polymer research and production is expected.

Perhaps the highest quality research in materials science is carried out at the Institute of Ceramic Chemistry and Technology in Shanghai and at the Institute of Metallurgy, which is located directly across the street; these two institutes apparently form the core of the basic materials R&D efforts in China. Research at the Institute of Ceramic Chemistry and Technology includes studies of single-crystal growth, such as inorganic and nonmetallic crystals including synthetic diamonds, work on glass ceramic materials, ceramics for electronics, high-temperature ceramics (in particular, oxides, nitrides, and carbides, including silicon nitrite), and ion-conducting ceramics. Much of the work at this institute is for long-range needs of electronic and laser applications. The institute in the past has carried out some work on carbon fibers as well as preliminary work on optical fibers. In addition to the Institute of Metallurgy in Shanghai, high-quality work on metals and alloys is also supposed to be in progress at the Institute of Chemical Engineering and Metallurgy in Peking. Materials work at the Institute of Physics in Peking includes studies of amorphous magnetic materials and thin films produced by sputtering, materials for magnetic bubble memories, microwave ferrite materials used as isolators in microwave systems, and single crystals for use in lasers including YAG. In the high-pressure laboratory at the Institute of Physics artificial diamonds are being produced by both static and dynamic methods.

Institute of Ceramic Chemistry and Technology, Shanghai

The overall role and function of this institute appears to be most parallel to the Materials Research Institutes established by ARPA and currently funded by the National Science Foundation in the United States. The principal areas of research include:

Single crystal growth. Efforts in this area include the study of inorganic and nonmetallic crystals including synthetic diamonds. The institute is engaged in hydrothermal methods of growing crystals such as quartz and of growing crystals directly from the melt. Crystals grown directly from the melt include those useful for nonlinear optic, electro-optical, and acousto-optical applications. Crystals include, for example, lithium niobate, lithium tantalate, lead molybdate, and bismuth germanate.

Special glasses. Work on glass ceramic materials includes "crystallized glasses," pioneered by Stockey of Corning Glass Works. Such glasses are apparently widely investigated internationally, since they are normally much stronger than ordinary glass and their electro-optical properties are also unique.

Ceramics for electronics. Activity at the institute includes work on ferro-electric and piezo-electric ceramics and a ferroelectric material made transparent by a sintering technique.

High-temperature ceramics, in particular, oxides, nitrides, and carbides, including silicon nitride.

Ion-conducting ceramics, such as β-alumina and doped zirconium oxides. Work in the past has included studies of rare earth oxides, but for the time being work is limited to rare earths only as doping materials.

Visits at this institute were carried out in the following laboratories:

Ferrite laboratories. The focus activity in this lab was on producing ferrite materials by sintering. The product displayed was a mixed crystal of lead, lanthanum, zirconium, and titanium (PLZT). One sample was of the following composition: $Ph_{0.9}La_{0.1}Zr_{0.65}Ti_{0.35}O_3$. Excess lead oxide is evidently added to help form a liquid phase to aid sintering in an oxygen atmosphere.

Electric birefringence. We were shown a demonstration of electrically induced birefringence in a transparent "PLZT" sample. The demonstration involved two TV cameras focused from slightly different angles on an object in the laboratory. The images from these two monitors were superimposed on the same TV screen. A pair of glasses made with the ferroelectric material inserted between crossed-Nichols were then used to observe the TV screen. A pulsed square wave modulation at 30 Hz was used to switch from one camera to the other camera and from one of the lenses to the other, producing quite a good stereo image for the observer. Such induced birefringence is useful, for example, in the production of instrumentation for the study of circular dichroism.

Crystal growth laboratory. Lithium niobate crystals with various doping elements were being used for producing plates that could be used for holographic storage purposes. Other applications included a 30-MHz filter and a structure for providing up to 6-nsec delay for 30-MHz signals. Lead molybdate was also being produced for acousto-optical crystal applications. Equipment included banks of furnaces for continuous crystal growing.

Ceramics laboratory. We were shown a variety of samples of silicon nitride materials produced both by sintering method and by a hot press method. The silicon nitride is pulverized to 2–3 μ before sintering. The various components included seals and fittings for pumps handling corrosive materials, such as acids, bases, and sludge.

Study of properties and applications of crystals. The crystals grown in the above laboratories are tested in a variety of ways, including an acousto-optical study. This involves the diffraction of He–Ne laser radiation through a lithium niobate crystal modulated at 64.7 MHz. The node pattern of the standing acoustical wave creates a diffraction of the laser beam that can be readily seen. The He–Ne laser is made in Shanghai.

Optical doubling system. A Chinese-made YAG laser was being used to provide pumping radiation at 1.06 μ for doubling by lithium niobate crystal to 0.53 μ. The efficiency of doubling depends on the quality of the crystal and in their applications is between 10%–20%. This system is used to test crystalline materials for optical quality and also symmetry of the unit cell. Such techniques for screening optical materials also represent the current state-of-the-art in the United States.

Vacuum evaporation equipment. Antireflection coatings of MgF_2 were being applied to crystals such as lead molybdate used in laser doubling applications.

Amorphous semiconductors. Chalconides were being used to develop glasses with electrical and optical storage properties. This group was trying to make memories for computer applications. We were shown silicon wafers with large arrays of solid-state elements deposited upon them, presumably in semiconducting glasses containing group 6 elements such as Te. We were shown a demonstration that such junctions could be used as binary elements in a memory device.

Holography experiment. A holographic image was created using the semiconductor chalconides we had seen previously. The image was produced by using a CW laser (4880Å) that was split to provide a beam that was passed through the object and a reference beam. The hologram was being read out with a He–Ne cw laser.

A study of ancient pottery. The crystalline glaze on an extremely rare and valuable sample of pottery from the Yuan period (approximately 1300–1400 A.D.) was being studied both by electron microscopic and optical microscopic techniques. This laboratory has about six optical microscopes of excellent quality including a Zeiss (Jena) phase contrast microscope.

The Institute in the past has carried out some work on carbon fibers. Some preliminary work on optical fibers is still in a primitive stage. Most of the work in this area will be done in industrial ministries including the Ministry of Post and Telecommunications.

This was one of the highest quality scientific institutions that we visited in China. It is an example of some of the best research in chemistry and materials science being done under an institutional label, which would not lead one to imagine either the high quality or substance of the work. The Institute of Ceramic Chemistry and Technology brings chemistry and chemically trained people very actively into the solid-state problems in a way that appears to be extremely productive and often does not occur in American research and educational institutions.

Institute of Physics, Peking

Work in the area of solid-state physics at the Institute of Physics in Peking is concentrated on the magnetism of amorphous materials; the study of gallium–cobalt thin films, which are produced by sputtering, magnetic bubbles; and ferrite materials. The entire effort appears to be related to memory devices and technology useful for communications and computers.

In the high pressure laboratory they are working on the synthesis of artificial diamonds and are also attempting the synthesis of new materials using high pressure. In a two-step compression they can achieve pressures of 10^5 atmospheres with a total force of 1200 tons. The compression elements are tungsten carbide in sections of spheres that are then compressed together to form a spherical unit.

C. Environment

Environmental concerns do not appear to have high priority in the overall Chinese national plan. This is particularly evident in heavily industrialized areas such as Fushun, where discharge of pollutants into the atmosphere and into the rivers, unacceptable by U.S. standards, does not appear to be a matter of major concern. Figure 42 shows the general condition of the atmosphere in the area close to the Fushun open pit coal mine, while Fig. 43 shows the discharge of raw

Figure 42. Atmosphere in the Fushun area.

Figure 43. Discharge of raw NO_2 into atmosphere in Lanchow.

NO_2 into the atmosphere from a plant on the outskirts of Lanchow. In spite of serious pollution problems China has begun to show concern about environmental protection. This work is coordinated nationally by an Environmental Protection Agency, which was established in the early 1970s directly under the State Council.

The following quotation, taken from the Hsin Hua News Agency News Bulletin (Thursday, May 18, 1978), provides some insight into the status of environmental programs in China:

> Peking, May 17 (Hsinhua)—In a recent interview with Hsinhua a leading member at the office for environmental protection, which is run within the State Council, answered questions about the work and problems of environmental protection in China. He was also asked about the policies and measures being implemented by the office.
>
> Q. What is the state of environmental protection in China?
>
> A. Chairman Mao, Premier Chou En-Lai and Chairman Hua Kuo-Feng have paid great attention to environmental protection. Much has been done to improve and protect environmental conditions over the past 28 years since the founding of new China. Older cities have been transformed, conditions in workers' living quarters and in public health conditions have been improved. There is better distribution of industry. Small enterprises have merged with bigger ones, which have been moved to new industrial districts where multi-purpose use is made of wastes.
>
> Soon after the convocation of the first meeting on environmental protection held by the state council in 1973, protection groups were set up in all provinces, municipalities and autonomous regions and departments under the state council. They have carried out a general survey of air and water pollution in China and adopted effective control measures.
>
> New techniques and technological processes have been introduced to check pollution. They include apparatus that no longer requires the use of mercury, non-cyanide electroplating, enzyme processes (removing the hair from hides through the use of enzymes), ammonia base sulphite pulping, and the re-use of treated waste water at oilfields. In addition some factories have developed advanced techniques for the disposal of waste water, gas and slag.
>
> Q. How bad is environmental pollution in China?
>
> A. Environmental protection was seriously disrupted by the sabotage of the Gang of Four. Pollution in some cities and contamination of rivers, lakes and seas, in soils and in the working environment has not yet been controlled.

It is rather serious in some localities and industries. Density of dust and sulphur dioxide has gone beyond the limits set by the state. Major waterways such as the Yangtze, the Yellow, the Huaiho, and the Pearl have been seriously contaminated in sections running through industrial cities because untreated industrial waste liquids are discharged into them directly or indirectly. Contamination of Pohai, China's biggest inland sea, is becoming serious.

The amount of industrial waste is estimated at 200 million tons every year. Most of the waste is not utilized, and takes up space and pollutes the environment.

Noise pollution in many cities particularly in mining or industrial areas is also very serious.

With industrial development, environmental pollution is now given even greater attention.

Q. What further measures are you going to take to protect the environment?

A. A mass movement is needed to speed up pollution control.

The anti-pollution projects will be put into the state plan so as to tackle industrial pollution in stages and according to the order of importance and urgency. The funds and materials needed will be guaranteed. Industries causing pollution must take realistic measures to solve them and meet environmental standards set.

Industries that are new or involved in building projects should design and put into operation pollution control systems at the same time as building the main project. Otherwise they will not be allowed to build or operate. There needs to be careful checking and supervision that these regulations are met.

Work on the law and regulations for environmental protection will be completed soon.

We shall step up research on the environment and on techniques of controlling pollution. Particularly important are the finding of new techniques, multi-purpose utilization, all-round prevention and control, environmental analysis, as well as the development of monitoring skills and the mastering of basic theories of environmental science.

State policy is to encourage industry to turn industrial wastes into useful things. However, it is often the case that a single industry or unit cannot manage multi-purpose utilization. Joint or cooperative efforts are sometimes required.

The work of environmental protection is new to us. We have to publicize its importance on a grand scale and by making clear its significance to everyone, inspire the masses to greater efforts to fight pollution.

Institute of Environmental Chemistry, Peking

The Institute of Environmental Chemistry in Peking appears to have a leading role in research in the environmental field. A major activity at this Institute is the development of instruments (in cooperation with other institutes and manufacturers), analyzing samples, and data processing. The instrument developments have provided an automatic SO_2 monitor, an ozone monitor, a small-size graphite atomizer (pyrographic tube) for atomic absorption, and a coulometric instrument. The atomic absorption apparatus has been demonstrated on rare earths with a sensivitity to 0.1 parts per billion. Analyses are normally made for phenol, CN^-, Cr, Hg, As, Se, Ba, Cu, Zn, Cd, Pb, O_3, SO_2, NO_2 total NOX, and organics. They have also developed a method for the chemical oxygen demand (COD) for natural water and waste water with ceric ion as the oxidant in excess and back titration with ferrous ion.

An air pollution monitoring van is equipped for analyses for the purposes of hygiene and environmental protection in Peking. A major air quality problem in Peking is caused by dust from the Gobi Desert.

Futan University

A laboratory at Futan University carries out research and instrumental development in the environmental field and is the center for environmental and water pollution control in the city of Shanghai. They collaborate closely with the municipal water supply system. Shanghai city water contains 10 parts per billion of dissolved Pb^{+2} and less than 0.4 parts per billion of Cd^{+2}.

Laboratory Safety

The Chinese appear to have little concern about safety practices that are taken for granted in the United States. For example, it is rare to see a gas cylinder chained or safety showers. Frequently apparatus is haphazardly wired, or hot wires just dangle in the laboratory. Hoods in China appear to have doors made from windowpane glass. Eye protection appears to be nonexistent, not only in chemical laboratories, but also in machine shops and areas where laser beams or arcs are in use as well. In refineries one finds little evidence of most common safety practices, including appropriate handling of effluents, protective headcover, and hard shoes. Current priorities in China appear to give far greater emphasis to creating the industrial base and infrastructure required for rapid development than to issues of safety and environmental quality. This is consistent with the pattern typical in many developing countries. The Chinese clearly face major problems in these areas that will have to be resolved as part of their long-range development strategy.

8. INFRASTRUCTURE

A. Scientific Communications

While technical services, meetings, and journals operated before the Cultural Revolution, activities that were suspended during the period of 1966–72 are gradually resuming. The best work in all science and engineering areas, including chemistry and chemical engineering, is likely to be published in *Scientica Sinica*, which has an English language edition. *Science Bulletin* (Kexue Tongbao) (covering all branches of science) resumed publication in 1973 as did *Chemistry Communications* (Huaxue Tongbao). The *Journal of Analytical Chemistry* resumed in 1974. *Acta Chimica Sinica* resumed publication in 1975, and *Polymer Communications* resumed publication in October 1978. A chemical engineering journal was published before the Cultural Revolution, and it is hoped that publication will resume. An abstracting journal, *Problems of Chemistry and Chemical Engineering in China* (Chung-kuo Hua-hsueh Hua-kung Wen-ti), is published by the Institute for Scientific Information. This abstracts only important foreign journals and is intended for Chinese readers who do not understand English.

The Institute for Scientific Information carries out research on information technology and hopes to set up a national information network. It maintains liaison with the central library of the Academy of Sciences in Peking as well as with libraries of the various institutes. The Institute of Chemistry in Peking has been designated as the lead agency in developing a chemical information system for the entire country. It is doing this in conjunction with the Institute for Scientific Information, the Academy Library, and the National Library.

Libraries visited in research institutes appeared to have a good selection of Western journals. Collections were fairly complete and did not seem to have been interrupted during the Cultural Revolution. These included *Chemical Abstracts*, *Science*, and *Chemical & Engineering News*. Book collections (Western) were poor. Many of the books were outdated and recently published books were scarce. Patent literature was said to be available in microfilm form. Xerox-type duplicating machines were not available and copying was done by photostating. Interlibrary exchange of photocopies is possible.

Since the Chinese do not subscribe to international copyright conventions, they can make the international scientific literature available on quite an inexpensive basis; they take full advantage of this opportunity. The Chinese regularly receive a subscription to all of the significant scientific journals; each is then duplicated and widely distributed among the institutions of the country. However, the journals were often not easily recognized since the duplicated versions were bound in unusual covers, as shown in Fig. 44. In addition, the original journal from which these copies are made is also assigned to one of the libraries of the country. This duplication process leads to some time lag; in general, we found that the journals on the shelves were received anywhere from three to six months later than the indicated publication date.

Apparently these journals are widely read, at least among the top chemists in the country. We were repeatedly struck by the very good acquaintance the Chinese had with the U.S. chemical literature; they were aware of the major developments and knew of the leading figures. However, the group involved in such readership is relatively

Figure 44. Reproduced version of Chemistry & Engineering News *in Chinese library.*

small. By one estimate we received, approximately 15,000 Chinese scientists read at least on an occasional basis U.S. journals such as the *Journal of the American Chemical Society*. Presumably with the reinstitution of graduate programs the readership will increase. In the library of Peking University, for instance, there is a "Faculty Reading Room" in which the journals are kept. These are not available to students except on a special basis, but the Chinese agreed with our assessment that the reinstitution of graduate programs would have to involve making these journals accessible to the students. Books seem to be less available and the time lag may be two to three years. Some of the more popular books are duplicated, but there is apparently some sensitivity to the fact that international copyright conventions are not honored.

The chemistry libraries overall are quite substantial. As one of the better examples, Futan University in Shanghai has a chemistry library that is supposed to contain more than 20,000 books and to carry subscriptions to 800 journals published in 17 languages. The library is well organized with good working space and has a separate working room for the use of *Chemical Abstracts*.

By contrast, U.S. awareness of recent advances in Chinese chemistry is quite poor. The Chinese seem to have a rather strong policy against publication in foreign journals, and this extends even to international journals that do not have an identification with a particular country. Few scientists in China thought that they would have opportunity to publish their work in *Tetrahedron*, for instance, even though one of the regional editors of this journal is Huang Min-lan, a Shanghai chemist. On the other hand, the important advances in Chinese chemistry are available to Western readers in a few key journals. Of these the principal one is *Scientia Sinica*, which is published in both a Chinese- and an English-language version. This journal is apparently seen as the place to publish work that is important enough to be called to the attention of the outside world. In it are regularly reported, for instance, the structures of the active compounds that are being isolated and identified from the systematic exploration of Chinese traditional medicines. The pharmacological effects of these compounds are apparently often reported in the *Chinese Medical Journal*, which also is published with an English-language version. Since some of the work of greatest novelty being done in China now is the identification and medical assessment of a variety of new compounds contained in Chinese herbal medicines and produced by local strains of microorganisms, it would seem that a regular examination of this literature would be of great interest to U.S. medicinal and organic chemists. In addition, much chemical work of general interest is published in *Acta Chemica Sinica*, which is in Chinese only but which contains an English language abstract. In the area of natural product structure determination, the abstract along with the structural formulas is frequently sufficient to convey much of the essential information. Another journal of this type, in which some of these natural product isolations are reported, is *Acta Botanica Sinica*. Again, this is in Chinese with an English abstract. As Chinese chemistry begins to gather strength after its 10-year hiatus, we may expect that awareness of this literature will be even more important to Western scientists, and we may hope that the Chinese will begin to participate in publication in international journals in their field at a significant rate and cooperate through involvement with *Chemical Abstracts*.

Scientific societies also discontinued activities during the Cultural Revolution and are just in the process of resuming operations. There is a single chemical society and not separate societies for subfields like polymer chemistry. The first Quantum Chemical Symposium was held in September 1977, and a second symposium is planned for 1979. These meetings are organized by the Chinese and have largely Chinese participation (by invitation). However, there may be some international participation in next year's meeting. A general meeting of the Chemical Society was held in September 1978 in Shanghai with attendance by 500–600 people. A meeting on polymer physics was planned for 1978 and one on polymer chemistry for 1979. Internationally attended meetings are just resuming. A biologically oriented symposium sponsored jointly by the Chinese and Australians was held in Peking in May 1978. Representatives of nine countries attended.

It is not difficult for qualified Chinese scientists to attend domestic meetings. However, attendance at foreign meetings is just resuming. Six nuclear scientists attended the Gordon Research Conference on Nuclear Chemistry in New Hampshire in June 1978. This was possible because it coincided with the visit of a Chinese delegation of nuclear scientists to the United States. There is enthusiastic interest among Chinese scientists for attending future Gordon Conferences in other areas, and it would be worthwhile to extend invitations.

In addition to visits by delegations, visits by individual scientists to China are possible by special arrangement. For example, Professors H. Mark, S. Atlas, and H. Morawitz (Brooklyn Polytechnic), and Dr. J. Hwa (Stauffer) have visited previously. Dr. L. Tung (Dow) visited in 1978 and Professor P. J. Flory (Stanford) visited for two weeks in September and October 1978. There have not yet been corresponding visits by individual Chinese scientists to the United States. Prior to the fall of 1978 there were also no instances of Chinese scientists serving as visiting scientists or postdoctoral fellows.

B. Scientific Manpower

Most of the leading scientists in China who are from 55–70 years of age are Western-trained, with many having received their advanced degrees in the United States. It is this group that appears to have emerged from the Cultural Revolution and the Gang of Four to resume the leadership of the major research institutions in China. A second quite distinct cohort of scientific manpower is the group in the age range from 45–55, most of whom were trained in China and never went abroad, and some of whom were trained in the Soviet Union in the period between 1950–60. In general, it is more difficult to find within this group enough people with the advanced training appropriate to provide the next generation of leadership. Although we met some highly qualified scientists who were trained in China from 1960 to the present, the number of people trained, particularly during the Cultural Revolution and the Gang of Four, was very small. In the universities, academic institutes, and the laboratories of the industrial ministries, there are very few active young investigators in the age range from 30–45.

The effect of this missing cohort of scientific and technical manpower is bound to be a serious impediment to the Chinese in achieving both their short- and long-term development objectives. There is every indication that the Chinese understand this problem and will be extremely active in trying to repair the damage by admitting students throughout this age group to a variety of advanced programs in both universities and research institutes in China, as well as by greatly increasing the emphasis on sending students abroad for specialized training and for participation in degree programs.

C. Innovation

In China there appears to be little attempt to innovate in the sense of developing entirely new technology based on indigenous fundamental research. However, there is considerable effort to introduce known technology into Chinese organizations, and the pace is exceedingly rapid at some centers such as oil refineries and petrochemical complexes, which must have severely strained the limited science and engineering manpower. The effort required for an innovation should not be underestimated, even when it involves the introduction of known technology with outside help. In the United States, when a corporation licenses a process from another corporation, considerable effort and ingenuity are still needed to adapt the process that has already been tested to the new organization.

Many industrial catalysts in China are said to be self-developed and manufactured including zeolite cracking catalysts and multimetal refining catalysts. Many of the older processes for inorganic chemicals such as nitric acid and caustic soda are apparently self-developed, based on existing Western models. In the Shanghai Petrochemical Complex, a new idea that has been discussed in the United States for a long time but never put into practice has been tried, namely a petrochemical refinery where oil is processed mainly for petrochemical use rather than for fuel use. This refinery contains many brand new processes that are imported from Japan and West Germany and built with Japanese help.

In the spread of new technology from a center of innovation or adoption to other places, the Chinese have demonstrated considerable skill. A good example of this process involves the refineries at Lanchow and Yumen. These were the first refineries in China after Liberation and therefore needed to be complete refineries to supply all the needs of the vast country. Very many of the Chinese refinery scientists and engineers received their first practical training and operating experience here. As new oil has been discovered, the newer refineries have tended to be less complex, because there was less need for them to be complete and self-sufficient. The refinery at Lanchow thus became a place where a good deal of the leadership of the new refineries was developed and has become a training center of manpower and a testing ground for new technologies for the rest of China. The Lanchow refinery also manufactures fluid catalytic cracking catalysts not used there but needed at other refineries. Other examples of such centers for technology transfer can be found for the development of specific types of scientific instrumentation.

D. Resource Allocation

We investigated how research projects are initiated and how the funding is determined. In the case of a project related to an industrial need, the relevant industry can ask the Chinese Academy of Sciences to have the work done, and the Academy can then request this of the appropriate institute. The institute may or may not agree to take on the project, depending on the field of work involved and the extent of the commitments the institute already has. A significant fraction of the projects are initiated within the institute itself. A new idea can be proposed to the Institute Academic Committee (which is usually composed only of members from the institute itself). Typically this committee might include both senior scientific personnel and younger members, and the composition of the committee changes. This committee then decides whether the proposed idea should be pursued.

Most funding from the Academy of Sciences is provided to the institutes in block form. The Chinese assured us that their budgets increase every year and availability of funds is not a serious problem. However, there is a separate budget for equipment that must be bought overseas with hard currency. This budget is highly restricted and controlled directly by the Academy of Sciences in Peking. Competition for the limited amount of such hard currency funds seems to explain some of the instrumentation deficiencies that we saw in many institutes.

The process of authorization for the heavy ion accelerator at the Institute of Modern Physics in Lanchow represents an interesting example of how such decisions on major facilities are made. After the submission of the proposal for construction to the Chinese Academy of Sciences, a national *ad hoc* committee was appointed to review the proposal and make its recommendation. The committee included scientists and engineers from institutes and universities, representatives from industry, and a representative from the Science and Technology Committee of Kansu Province (where Lanchow is located). The endorsement of this Committee was apparently helpful in gaining final approval for the project.

9. DISCUSSION

The level of Chinese effort and achievement in chemistry and chemical engineering is substantial but uneven. There are areas of obvious high priority, but also some significant gaps.

An area of particular interest is the high-quality effort in pharmaceutical chemistry, particularly development of pharmaceuticals from traditional Chinese medicinal materials, which is represented by programs at the Institutes of Materia Medica in Peking and Shanghai and at the Institute of Organic Chemistry in Shanghai. Another significant project in organic chemistry is the yeast alanine *t*-RNA synthesis at the Institute of Biophysics in Shanghai.

Polymer chemistry is a more applied area of intensive Chinese effort. China produces most of the conventional polymers and has a broadly based research program with emphasis on applications and short-range goals. The strongest efforts focus on the development of new polymerization catalysts. Another applied area of considerable interest is the materials work at the Institute of Ceramic Chemistry and Technology in Shanghai on the development of solid-state materials for electronic, electro-optical, and acousto-optical use, including piezoelectric materials. Chinese work in gas chromatography is extensive and uniformly good.

China is developing an extensive and modern petroleum and petrochemical industry that includes the required infrastructure and research base. One area of particular interest is China's experience in the operation of full-scale shale oil production facilities. The Chinese effort in catalysis is extensive and appears to be well-coordinated with their development of chemical process industries. The most obvious deficiency is the lack of on-line analytical techniques and microprocessors for process control.

In contrast to these areas of strength, modern organometallic chemistry appears to be practically nonexistent in China. Although Chinese organic chemistry is strong in some areas, there has not been enough emphasis on the invention of new synthetic reactions, on the invention of new methods for structure determination, or on the development of physical organic chemistry. There is also very little crystallographic work on automated equipment for X-ray diffraction studies. Furthermore, except for the strong program in molecular orbital theory, there is essentially no quantum and theoretical chemistry, statistical mechanics, or molecular spectroscopy—microwave, infrared, visible, UV, NMR, ESR—and no current program in molecular beam research and microscopic kinetics, although serious programs in these areas are planned.

Although there are computers in many of the research institutes in China, chemists do not always have easy access to them, and they have not been aggressive in developing applications for computers in their work, such as simulating spectra and fitting kinetics data. The computers that might be available are behind the state-of-the-art, but they are adequate for many chemical applications.

An explanation for these gaps may be that many of these areas were either started or advanced rapidly during the past decade, a decade in which the Chinese chemical community was preoccupied with very applied research and with political upheaval. Another reason is that work in many of these areas depends heavily on computational capabilities.

Chemical research in China today is highly application-oriented. It is mainly concerned with day-to-day problem-solving, generation of process information geared to improving existing processes, improvement in catalysts for existing processes, and development of improved products. There is little research aimed at understanding underlying phenomena or attempting to recast problems in terms of basic scientific questions. Also almost nonexistent are separately managed teams of scientists devoted to long-range research aimed at pioneering new products or new processes. It is evident that the Chinese have decided at least up until now to emphasize the short-range and sacrifice the long-range in order to maximize the short-term development of their productive capacity. There is much interest in how the United States conducts its long-range research and how it is meshed into existing programs. As China continues on its rapid industrialization program and as its scientists develop a better understanding of critical technical issues, the future may see increasing emphasis on longer-range research.

Scientific manpower will remain a critical issue to the future development of Chinese chemistry and chemical engineering. Today's leadership is generally in the hands of scientists who range from 55 to 70 years of age and who were trained in the West. Measures must be taken to retrain and upgrade the age cohort from 30 to 45 and to graduate larger numbers of well-trained chemists and chemical engineers. This will almost certainly require improved efforts to tie together graduate training in the universities and research in the academy and ministry institutes.

In areas of commercial interaction, although the Chinese have purchased U.S. technology and scientific instrumentation when there appeared to be no other real choice, they seem to have preferred to deal with countries that had already established formal diplomatic relations. In the field of instrumentation, for instance, we were struck by the preponderance of Japanese instruments among those that are imported. Even instruments designed in the United States were frequently purchased from foreign subsidiaries of U.S. companies, presumably in part to avoid various U.S. and CoCom export restrictions.

The Chinese are launching a major program to upgrade science and technology, and this will undoubtedly involve major purchases of scientific instrumentation from overseas. Furthermore, after a long period of inadequate scientific training, the Chinese are now interested in sending students and more advanced scientists overseas for further education. Considering the enormous goodwill that we sensed particularly among those Chinese who had studied in the United States and the establishment of formal diplomatic relations in December 1978, U.S. participation in the market for high-technology products and in the overseas training of Chinese students and research fellows should grow substantially.

With China increasingly turning its attention to possible interactions with the rest of the world, it is of interest to try to gauge the position that the United States might occupy relative to other nations. The United States has started with some enormous advantages in its interactions with China. All Chinese students must learn a second language, and English is the most popular. Far behind English as a second language are Japanese, Russian, German, and French. Chinese chemists regularly read and understand the English language scientific literature, and the major universities and research institutes carry a full selection of U.S. journals.

Another advantage we have is the fact that the Chinese have a very strong commitment to applied chemistry, and the United States is perceived to be the leading country in this area. Furthermore, a remarkably large fraction of the leadership of chemistry in China is U.S.-trained. This group retains a strong identification with our country and normally would encourage students and other Chinese scientists to study in the United States. Apparently even those Chinese chemists who received their training in the Soviet Union now share the general feeling in China that the Soviet Union is the chief threat to the security of their country. Thus, the reservoir of goodwill felt toward the United States and specifically the admiration of our basic and applied science is very large.

The major question for the future is whether the Chinese will be able to absorb and capitalize on high technology. The principal uncertainties are the chance for political stability. and the survival of a pro-science and technology political leadership. Political instability has been one of the chief factors historically in inhibiting the development of modern science and technology in China. It appears that China has made a decision to rely heavily on foreign technology as the source of innovation for economic development, but the leadership also clearly recognizes the importance of having indigenous R&D capability for effectively absorbing foreign technology. It remains to be seen whether China can create its own innovation chain from R& D to commercial implementation. Given the essential elements of political stability, good leadership, adequate resources, and a functioning infrastructure of research and training institutions, the results could be dramatic.

10. RECOMMENDATIONS

A mechanism for maintaining continuing contact with chemistry and chemical engineering in China should be explored. Various aspects of such a continuing relationship might include:

1. The attendance of Chinese scientists at key scientific meetings in the United States, and the attendance of U.S. scientists at meetings in China. Such meetings could include bilateral conferences in mutually agreed areas, Gordon Conferences, American Chemical Society meetings, AIChE meetings, and American Physical Society meetings (Division of Chemical Physics and High Polymer Physics, for example).
2. An exchange of Chinese and U.S. working scientists. This could include middle-level working scientists, postdoctoral fellows, graduate students, as well as special students, in various agreed-upon areas of technology.
3. A small number of joint scientific programs in substantive areas of interest to the Chinese as well as to the U.S. scientific community. Examples of such areas might include:
   (a) Natural products chemistry
   (b) Shale oil production engineering
   (c) Gas chromatography
   (d) Polymer physics and chemistry
   (e) *t*-RNA synthesis
   (f) Heavy-ion nuclear chemistry
   (g) Scientific information

APPENDIX A Science and Technology Organization in China

NATIONAL ORGANIZATION FOR SCIENCE AND TECHNOLOGY - FOCUS ON CHEMISTRY AND CHEMICAL ENGINEERING

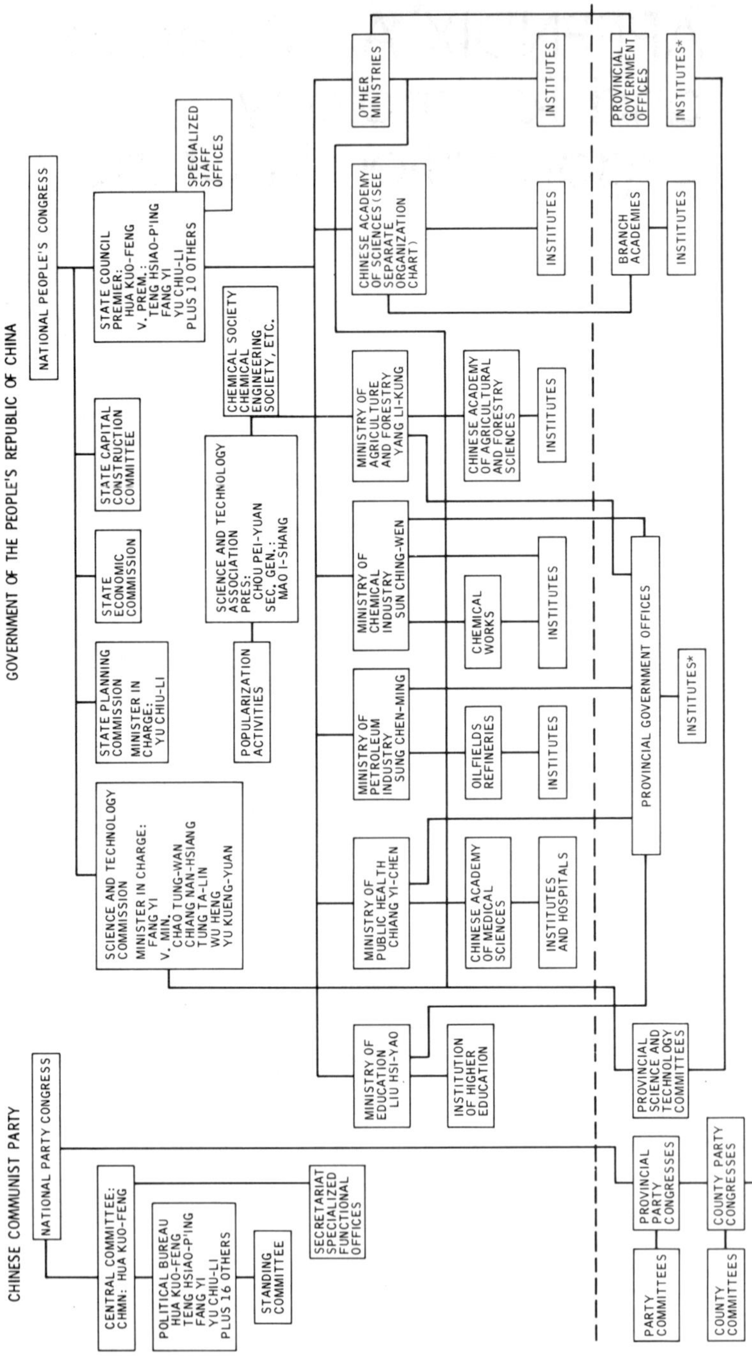

*SOME UNDER JOINT CENTER-LOCAL CONTROL

ORGANIZATION OF THE CHINESE ACADEMY OF SCIENCES - FOCUS ON CHEMISTRY

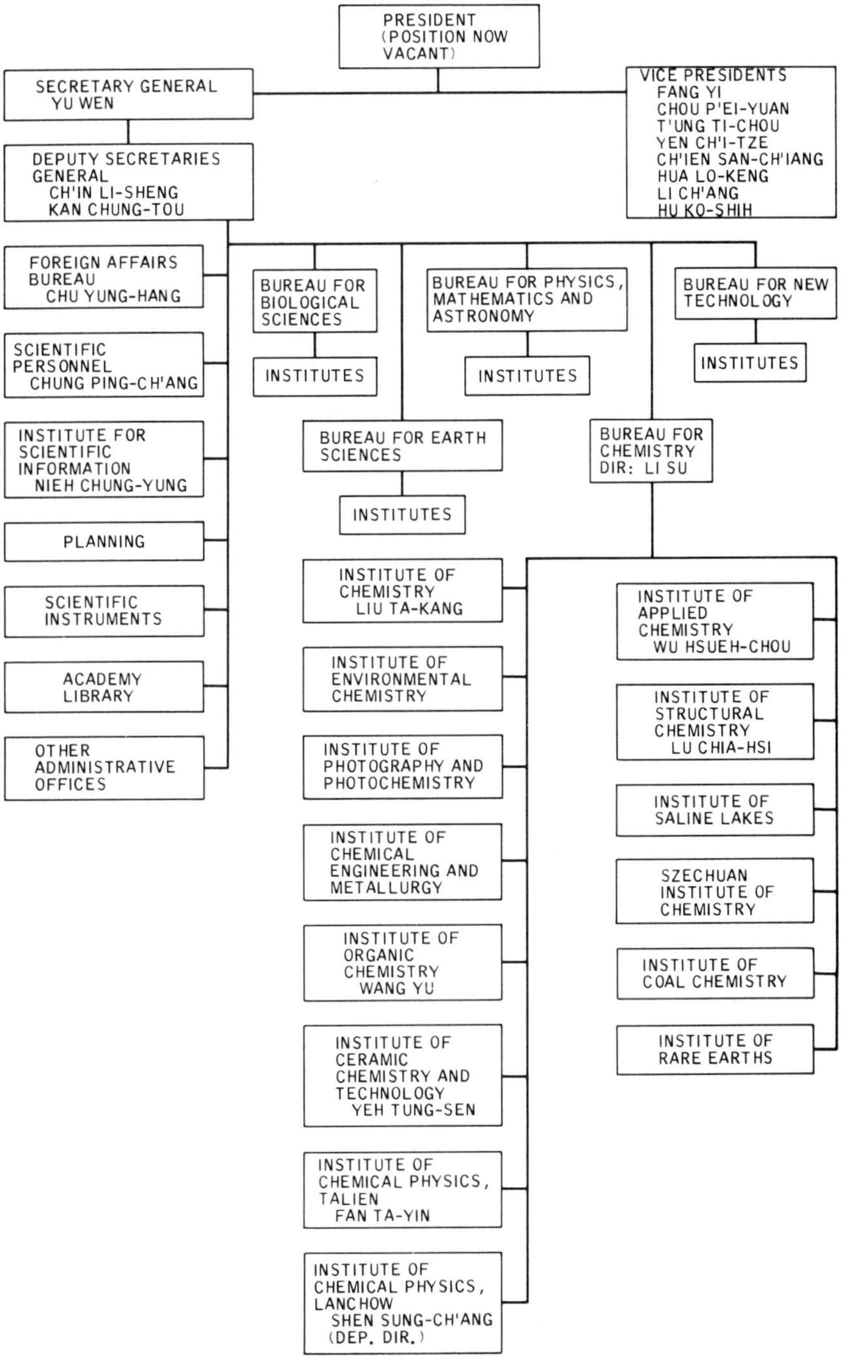

APPENDIX B Institutions Visited (in the order visited)

	Institute	*Location*
1.	Institute of Chemistry	Peking
2.	Institute of Physics	Peking
3.	Peking University	Peking
4.	Tsinghua University	Peking
5.	Institute of Atomic Energy	Peking
6.	Institute of Materia Medica	Peking
7.	Institute of Biophysics	Peking
8.	Petrochemical Research Institute	Peking
9.	Peking General Petrochemical Works	Peking
10.	Institute of Environmental Chemistry	Peking
11.	Institute of Photography*	Peking
12.	Peking Vinylon Plant	Peking
13.	Institute of Chemical Physics	Talien
14.	Talien New Harbor	Talien
15.	Talien Locomotive Factory	Talien
16.	No. 7 Oil Refinery	Talien
17.	Tach'ing Academy of Science, Technology, and Design	Tach'ing
18.	Tach'ing General Petrochemical Works	Tach'ing
19.	Research Institute of Tach'ing General Petrochemical Works	Tach'ing
20.	Institute of Petrochemistry	Harbin
21.	Kirin University	Ch'angch'un
22.	Kirin Institute of Applied Chemistry	Ch'angch'un
23.	New China Printing Plant	Ch'angch'un
24.	Refinery Complex No. 2	Fushun
25.	West Open Pit Coal Mine	Fushun
26.	Institute of Organic Chemistry	Shanghai
27.	Institute of Biochemistry	Shanghai
28.	Institute of Nuclear Research	Shanghai
29.	Institute of Materia Medica	Shanghai
30.	Futan University	Shanghai
31.	Shanghai Institute of Chemical Engineering	Shanghai

* Not visited, but chemical work going on there discussed by hosts.

32. General Petrochemical Works	Shanghai
33. Institute of Ceramic Chemistry and Technology	Shanghai
34. Shanghai Normal University	Shanghai
35. Shanghai Water Works	Shanghai
36. Computer Technology Research Institute	Shanghai
37. Shanghai Industrial Exhibit	Shanghai
38. Shanghai Refinery*	Shanghai
39. Chekiang University, Department of Chemical Engineering	Hangchow
40. Northwest University, Department of Chemistry	Sian
41. Institute of Chemical Physics	Lanchow
42. Institute of Modern Physics	Lanchow
43. Academy of the Lanchow Chemical Industry	Lanchow
44. Lanchow Institute of Petroleum Research	Lanchow
45. Lanchow Petroleum Refinery	Lanchow
46. Liuchiahsia Hydroelectric Plant	Liuchiahsia, Kansu Province

* Not visited, but chemical work going on there discussed by hosts.

APPENDIX C Principal Delegation Escorts

In Peking:

胡凤仙	*HU Feng-hsien (f)	Ministry of Foreign Affairs
史维明	*SHIH Wei-ming	Peking Institute of Chemistry
曹旋旋	*TS'AO Hsuan-hsuan (f)	Peking Institute of Environmental Chemistry
張小晓	*CHANG Hsiao-hsiao (f)	Division of Foreign Affairs, Chinese Academy of Sciences
邱秉钧	*CH'IU Ping-chün	Division of Foreign Affairs, Chinese Academy of Sciences

On the road:

钱人元	*CH'IEN Jen-yuan	Deputy Director, Peking Institute of Chemistry
施良和	*SHIH Liang-ho	Professor, Peking Institute of Chemistry
邓少林	TENG Shao-lin	Librarian, Peking Institute of Chemistry
史维明	SHIH Wei-ming	
胡凤仙	HU Feng-hsien (f)	

* Met at Peking Airport, May 17, 1978

NOTE: (f) on this and following pages indicates female.

Host Institution Name Lists

1. Institute of Chemistry, Peking—May 18, 1978

柳大纲	*LIU Ta-kang	Director
钱人元	*CH'IEN Jen-yuan	Deputy Director and Delegation Escort
胡亚东	*HU Ya-tung	Secretary General and Associate Research Fellow
王葆仁	WANG Pao-jen	Professor
施良和	*SHIH Liang-ho	Professor and Delegation Escort
崔孟元	TS'UI Meng-yuan	Lecturer, "Photo-oxidation and Photo-degradation of Polypropylenes"
黄志镗	HUANG Chih-t'ang	Lecturer, "Heterocyclic Polymers"
蒋明谦	CHIANG Ming-ch'ien	Professor, "Quantitative Relationship between Molecular Structure and Physico-Chemical Properties"
严继民	YEN Chi-min	Lecturer, "Graph Theory of Molecular Orbitals"
叶成	YEH Ch'eng	Lecturer, "Organic Conductors (TTF and TCNQ)"
吴培基	WU P'ei-chi	Lecturer, "Holography"
邓礼如	TENG Li-ju (f)	Lecturer, "Polymeric Packing Materials for Chromatography"
王守道	WANG Shou-tao	Lecturer, "Crystal Structure of Monocrotaline"
徐懋	*HSU Mao	Lecturer, "Polymeric Crystals"
史维明	*SHIH Wei-ming	Office of General Affairs and Delegation Escort

*Met at Peking Airport, May 17

	YEN Hai-ko	Thermochemistry Laboratory
	PIAN Tzu-liang	Mass Spectrometry Laboratory
	LI Chung-ming	Mass Spectrometry Laboratory
陈观文	CH'EN Kuan-wen	Staff
马福荣	MA Fu-jung (f)	Staff
邓少林	TENG Shao-lin	Librarian and Delegation Escort

2. Institute of Physics, Peking—May 18, 1978

管维炎	KUAN Wei-yen	Deputy Director
王汝敬	WANG Ju-ching	Foreign Affairs Section; Liquid Helium Laboratory
韩建国	HAN Chien-kuo	Interpreter
谢章先	HSIEH Chang-hsien	Staff
李银安	LI Yin-an	Fusion Laboratory Briefer
刘家瑞	LIU Chia-jui	Tokamak Laboratory Briefer
沈主同	SHEN Chu-t'ung	Interpreter and Responsible Person in Fusion Labs
陈祖德	CH'EN Tsu-te	Interpreter and Responsible Person in Fusion Labs
	SHU Chi-jen	Laser Laboratory
	LIU Yin-lieh	Magnetic Bubble Laboratory
	MO Yü-chun	Crystal Displays

Hosts, Welcome Banquet for the Delegation, May 18, 1978

周培源	CHOU P'ei-yuan	Director, Scientific and Technical Association; Vice President, CAS (Chief Host)
李苏	LI Su	Chief, Chemistry Division, CAS

朱永行	*CHU Yung-hang	Deputy Director, Division of Foreign Affairs, CAS
冯因复	FENG Yin-fu	Division of Foreign Affairs, CAS
李明德	LI Ming-te	Division of Foreign Affairs, CAS
张小晓	CHANG Hsiao-hsiao (f)	Division of Foreign Affairs, CAS
邱秉钧	CH'IU Ping-chün	Division of Foreign Affairs, CAS
胡凤仙	HU Feng-hsien (f)	Ministry of Foreign Affairs
廉正保	*LIEN Cheng-pao	Ministry of Foreign Affairs
肖心文	*HSIAO Hsin-wen	Ministry of Foreign Affairs
柳大纲	LIU Ta-kang	Director, Institute of Chemistry, Peking
钱人元	CH'IEN Jen-yuan	Deputy Director, Institute of Chemistry, Peking
胡亚东	HU Ya-tung	Secretary General, Institute of Chemistry, Peking
郭础	KUO Ch'u	Institute of Chemistry, Peking
施良和	SHIH Liang-ho	Institute of Chemistry, Peking
陈观文	CH'EN Kuan-wen	Institute of Chemistry, Peking
徐懋	HSU Mao	Institute of Chemistry, Peking
邓少林	TENG Shao-lin	Institute of Chemistry, Peking
王葆仁	WANG Pao-jen	Institute of Chemistry, Peking
史维明	SHIH Wei-ming	Institute of Chemistry, Peking
张青莲	CHANG Ch'ing-lien	Responsible Person, Department of Chemistry; Professor of Chemistry, Peking University

* Did not attend the return banquet on May 22

吴金城	WU Chin-ch'eng	Chief Engineer, Institute of Petrochemical Research
汪德熙	WANG Te-hsi	Director, Institute of Atomic Energy
黄量	HUANG Liang	Director, Institute of Materia Medica
申葆成	SHEN Pao-ch'eng	Director, Institute of Environmental Chemistry
曹旋旋	TS'AO Hsuan-hsuan (f)	Institute of Environmental Chemistry
李观华	LI Kuan-hua	Professor of Chemistry, Inner Mongolia Teachers' College

3. Peking University, Peking—May 19, 1978

张龙翔	CHANG Lung-hsiang	Dean of Studies
张青莲	CHANG Ch'ing-lien	Responsible Person, Department of Chemistry; Professor of Chemistry
徐振亚	HSU Chen-ya (f)	Responsible Person, Department of Chemistry
黄子卿	HUANG Tzu-ch'ing	Professor of Physical Chemistry
唐有祺	T'ANG Yu-ch'i	Professor of Physical Chemistry
徐光宪	HSU Kuang-hsien	Professor of Physical Chemistry
高小霞	KAO Hsiao-hsia (f)	Professor of Analytical Chemistry
冯新德	FENG Hsin-te	Professor of Polymer Chemistry
邢其毅	HSING Ch'i-yi	Professor of Organic Chemistry
顾孝诚	KU Hsiao-ch'eng	
	CHO Ching	
	CHANG Ÿ-ming	Student of T'ANG Yu-ch'i

4. Tsinghua University, Peking—May 19, 1978

	CHANG Wei	Vice-President (not met)
汪家鼎	WANG Chia-ting	Chairman, Department of Chemical Engineering
周昕	CHOU Hsin	Professor of Inorganic Chemistry
周其庠	CHOU Ch'i-hsiang	Polymer Chemist
彭秉璞	P'ENG Ping-p'u	Associate Professor
马文中	MA Wen-chung	Responsible Person, Office of the President
	WANG Shu-liang	Electron Scanning Microscope
	LIU Mi-chin	Mass Spectrometer
	CHIN Yun	Chemical Engineering Laboratory

5. Institute of Atomic Energy, Peking—May 20 and 22, 1978

钱三强	CH'IEN San-ch'iang	Director (now inactive)
汪德熙	WANG Te-hsi	Deputy Director
	LI Shen-nan	Deputy Director
	YANG Ming-chan	Chief of the Director's Office
	CHO Yi-chung	Deputy Chief, Theoretical Division
	HO Chang-yü	Deputy Director, Radiochemistry Laboratory
	LO Shang-kan	Research Associate
	YU Cheng-tse	Head, Center for Waste Treatment
	CH'EN Pai-sun	Research Assistant
	HSU Yuan-chao	Research Assistant
	WANG Shih-hsiu	
	CHANG Hsi-chen	Interpreter
	HSIAO Yi-chung	Chief, Radioisotope Group
	LIN Chen-kuo	Vice-Chief, Engineering Reactor Laboratory
	CHANG Wei-ming	Electromagnetic Separation
李观华	LI Kuan-hua	Professor of Chemistry, Inner Mongolia Teachers' College

6. Institute of Materia Medica, Peking—May 20, 1978

黄量	HUANG Liang	Director
	TU P'in-ch'eng	Director of Research
	FU Kui-hsiang	
	YAO Pu-chen	

7. Institute of Biophysics, Peking—May 20, 1978

	TSOU Chen-lu	Director, Molecular Biology Activities

8. Petrochemical Research Institute, Peking—May 20, 1978

	LU Cheng-ch'iu	Director
	CH'EN Chu-p'i	Vice-Director, Planning
	WANG Hsueh-chin	Chief, Analytic Chemistry Division
吴金成	WU Chin-ch'eng	Chief Engineer
	HU Ching-cheng	General Engineer and Escort
	CH'ENG Ch'ih-kuang	Vice-Chief Engineer
	WU Pao-hsun	Vice-Chief Engineer
	LIN Chu-ch'ao	
	HSIEH An-hui	
	LI Shu-hsun	
	CHAO Hsueh-pin	
	LI Sung-nien	

9. Peking General Petrochemical Works, Peking—May 20, 1978

傅叔勉	FU Shu-mien	Responsible Person, Department of Engineering Technology
陈志英	CH'EN Chih-ying	Chief, Office of Public Relations
石国柱	SHIH Kuo-chu	Staff, Office of Public Relations
刘进昌	LIU Chin-ch'ang	Interpreter, Foreign Affairs Office

10. Institute of Environmental Chemistry, Peking—May 20, 22, 1978

申葆成	SHEN Pao-ch'eng	Director and Head, Academic Committee
	CH'EN Chia-ho	Chief, Division of Scientific Administration
曹旋旋	TS'AO Hsuan-hsuan (f)	

12. Peking Vinylon Plant, Peking—May 22, 1978

杨振仓	YANG Chen-ts'ang	Deputy Plant Manager
孙国梁	SUN Kuo-liang	Chief, Plant Office
水佑人	SHUI Yu-jen	Engineer
牛良玉	NIU Liang-yü	Chief, Technical Section
陈心伦	CH'EN Hsin-lun	Deputy Chief, Vinylon Workshop
李宜忠	LI Yi-chung	Deputy Chief, (Research) Laboratory

Talien—May 23–26, 1978

曹林	*TS'AO Lin	Chairman, Talien Science and Technology Committee (Banquet Host, May 24)
寇根生	*K'OU Ken-sheng	Deputy Chairman, Talien Science and Technology Committee
孟宪武	*MENG Hsien-wu	Talien Science and Technology Committee
	*SUI Ta-yung	Talien Science and Technology Committee
孙云	*SUN Yun	Talien City Foreign Affairs Bureau

* Attended banquet, May 24

13. Institute of Chemical Physics, Talien—May 24, 1978

范大因	*FAN Ta-yin	Director
顾以健	*KU Yi-chien	Responsible Person, Scientific and Technology Department, Associate Researcher
姜炳南	*CHIANG Ping-nan	Deputy Director, Research Division
张乐沣	*CHANG Lo-feng	Deputy Director, Research Division
梁东白	LIANG Tung-pai	Assistant Researcher and Interpreter
姜仁鉴	CHIANG Jen-chien	Staff
宋延奎	SUNG Yen-k'ui	Staff
	WANG Hsieh-tao	Trace Analysis
	TSUNG Hsien-mou	Organometallic Compounds
	LIAO Shih-ch'ien	Olefin Polymerization
	TAO Lung-jan	Reaction Chemistry of Microcatalytic Conversion Processes
	HSIN Chin	Chemisorption
	SUN Yen-k'ui	High Molecular Weight Polycyclic Compounds
	LI Hai	Librarian
	LU P'ei-chang	Lecturer (May 25), "Some Theoretical Problems in the Development of Column Chromatography"; Medal Recipient at March 1978 National Science Conference

14. Talien New Harbor, Talien—May 24, 1978

MENG Yi	Manager, Harbor Authority
SHANG Chun-yin	Staff

* Attended banquet, May 24

15. Talien Locomotive Factory, Talien—May 25, 1978

	CHU Chung-cheng	Responsible Person

16. No. 7 Oil Refinery, Talien—May 25, 1978

	CHIN Hsi-kao	General Manager

Harbin

	CHANG Hen-hsuan	Deputy Secretary, Heilungkiang Party Committee; Vice-Chairman, Heilungkiang Province Science and Technology Committee (Banquet Host)
张明仁	CHANG Ming-jen	Office Vice-Director, Heilungkiang Province Science and Technology Committee
	LI Chun-hua	Heilungkiang Province Science and Technology Committee
江德祥	CHIANG Te-hsiang	Inorganic Chemist, Department of Chemistry, Heilungchiang University
周定	CHOU Ting	Associate Professor of Chemistry, Harbin Technological University
	TSO Ti (f)	Associate Professor of Chemistry, Harbin Technological University
	TU Chiao-pin	Deputy Chief Engineer, Harbin Bureau of Chemical Engineering

Tach'ing

崔海天	TS'UI Hai-t'ien	Vice Chairman, Tach'ing Revolutionary Committee
	MA Chan-ho	Tach'ing Bureau of Foreign Affairs

17. Tach'ing Academy of Science, Technology, and Design—May 24, 1978

	WANG Chuan-yü	Director
	WANG Chih-wu	Vice Director
	CHAO Tzu-hui	Chief, Office of Science and Technology
	KAO Wei-pao	Chief of Engineering
张有奎	CHANG Yu-k'ui	Research Scientist, Production Research Laboratory

18. Tach'ing General Petrochemical Works, Tach'ing—May 25, 1978

李革非	LI Ko-fei	Deputy Director
	WANG Ying	Deputy Chief Engineer; Vice-Chairman, Tach'ing Revolutionary Committee
	FU Cheng-tsung	Science and Technology Committee
	LIU Tung-t'ien	Research Office
	HUA Wen-yi	Reception Office

19. Research Institute of Tach'ing General Petrochemical Works, Tach'ing—May 25, 1978

	SUN Yi-ling (f)	Vice-Director and Chief Engineer
张景存	CHANG Ching-ts'un (f)	Chief of Exploration

20. Institute of Petrochemistry, Harbin—May 26, 1978

宋九福	SUNG Chiu-fu	Deputy Director
张淑芝	CHANG Shu-chih	Interpreter
	HSUEH Hai-yin	Chief, Administrative Office
	WANG Kuang-fu	Engineer and Chief of Research Activity
	WANG Tzu-lu	Inorganic Chemist (Adhesives); Associate Research Fellow
	LU Ch'i-ting	Adhesives Laboratory
	HSUN Hsi-wen	
	YEH Ts'ao-ch'ien	
	TS'AI Chu-ming	

Ch'angch'un—May 26–28, 1978

赫询	*HO Hsun	Deputy Director, Kirin Province Science and Technology Committee (Banquet Host, May 26)
韩圣谦	*HAN Sheng-ch'ien	Staff Kirin Province Science and Technology Committee
张志敏	*CHANG Chih-min (f)	Staff, Kirin Province Science and Technology Committee
辛连弟	*HSIN Lien-ti	Staff, Kirin Province Science and Technology Committee
刘日清	*LIU Jih-ch'ing	Kirin Province Bureau of Foreign Affairs

21. Kirin University, Ch'angch'un—May 26–27, 1978

唐敖庆	*T'ANG Ao-ch'ing	President
	KUO Feng-kao	Vice-Chairman, Revolutionary Committee
	TS'AI Lu-sun	Chairman, Department of Chemistry
江福康	CHIANG F'u-k'ang	Responsible Person, Department of Chemistry
丁德增	TING Te-tseng	Deputy Director, Division of Theoretical Chemistry
江元生	CHIANG Yuan-sheng	Professor, Department of Chemistry (graph theory of molecular orbitals)
孙家钟	SUN Chia-chung	Director, Division of Polymer Chemistry
沈家聪	SHEN Chia-ts'ung	Lecturer, Department of Chemistry
汤心颐	T'ANG Hsin-yi	Lecturer, Department of Chemistry (polymers)
罗修锦	LO Hsiu-chin	Lecturer, Department of Chemistry

* Attended banquet, May 26

金钦汉	CHIN Ch'ing-han	Lecturer, Department of Chemistry (molecular sieve)
徐如人	HSU Ju-jen	Lecturer, Department of Chemistry
岳贵春	YUEH Kui-ch'un	Lecturer, Department of Chemistry
于连生	YÜ Lien-sheng	Lecturer, Department of Chemistry
	CHANG Hung-an	Polymer Chemistry Laboratory
	HU Shih-yang	Semiconductor Department Laboratory
吴式枢	WU Shih-shu	Professor, Department of Physics
苟清泉	KOU Ch'ing-ch'üan	Professor, Department of Physics
杨善德	YANG Shan-te	Lecturer, Department of Physics
吴俊松	WU Chun-sung	Lecturer, Department of Physics
州介文	CHOU Chieh-wen	Lecturer, Department of Physics
王爱莲	WANG Ai-lien (f)	Lecturer, Department of Physics
舒文辉	SHU Wen	Lecturer, Department of Physics
刘運祚	LIU Yun-tso	Nuclear Physicist, Department of Physics
吴佩致	WU P'ei-tzu	Responsible Person, Office of the President

22. Kirin Institute of Applied Chemistry, Ch'angch'un—May 27, 1978

吴学周	*WU Hsueh-chou	Director and Research Fellow
钱保功	*CH'IEN Pao-kung	Deputy Director and Research Fellow

* Attended banquet, May 26

吴越	WU Yueh	Director, Catalysis Research Division; Associate Research Fellow
黄葆同	*HUANG Pao-t'ung	Director, Research Division; Associate Research Fellow (stereospecific polymerization)
汪尔康	*WANG Erh-k'ang	Associate Research Fellow (polarography)
張维钢	*CHANG Wei-kang	Associate Research Fellow and Interpreter
周恩乐	CHOU En-lo (f)	Assistant Research Fellow (polymer physics)
吴钦义	WU Ch'ing-yi	Associate Research Fellow (NMR)
沈联芳	SHEN Lien-fang	Deputy Chief, Structural Analysis; Assistant Research Fellow
	CHU Yü-fang	Briefer, Structural Analysis Laboratory
余赋生	YÜ Fu-sheng	Associate Research Fellow (polymer physics)
冯之榴	FENG Chih-liu (f)	Associate Research Fellow (polymer physics); wife of HUANG Pao-t'ung
姜炳政	CHIANG Ping-cheng	Assistant Research Fellow (polymer physics)
王佛松	WANG Fu-sung	Assistant Research Fellow (stereospecific polymerization)
姚克敏	YAO K'o-min	Assistant Research Fellow (laser chemistry)
徐纪平	HSU Chi-p'ing	Associate Research Fellow (polymer synthesis)
袁秀顺	YUAN Hsiu-shun	Chief, Inorganic Analysis Division; Associate Research Fellow

* Attended banquet, May 26

王文韵	WANG Wen-yun	Assistant Research Fellow, Laser Chemistry Research Division
周大凡	CHOU Ta-fan	Assistant Research Fellow, Laser Chemistry Research Division
林红	LIN Hung	Group Leader, X-Ray Research Division
杨振华	YANG Chen-hua	Researcher, Chemical Fluorescence
黄长常	*Huang Ch'ang-ch'ang	Staff
董绍俊	TUNG Shao-chün (f)	Rare Earth Metals, wife of WANG Erh-k'ang
苏锵	SU Ch'iang	Rare Earth Metals
倪加钻	NI Chia-tsuan	Lasers
相郁良	YANG Yü-liang	Lasers
李少宗	LI Shao-tsung	Solar Energy, husband of CHOU En-lo
杜有如	TU Yu-ju (f)	Solar Energy
	CHO Hsin-kung (f)	Interpreter

23. New China Printing Plant, Ch'angch'un—May 28, 1978

马宏录	MA Hung-lu	Deputy Plant Manager and Briefer
未菓生	WEI Kuo-sheng	Chief Administrator
李之明	LI Chih-p'eng	Chief, Planning Division
刘金福	LIU Chin-fu	Administrative Office

* Attended banquet, May 26

Shenyang/ Fushun—May 28–30, 1978

Shenyang Banquet (May 28) Hosts and Participants:

蔡藜	TS'AI Li	Deputy Director, Liaoning Province Science and Technology Committee (Chief Host)
纪万富	CHI Wan-fu	Office Chief, Liaoning Province Science and Technology Committee
马振国	MA Chen-kuo	Staff, Liaoning Province Science and Technology Committee
顾敬新	KU Ching-hsin	Chairman, Board of Directors, Liaoning Province Chemistry and Chemical Engineering Society
伍名俊	WU Ming-chün	Member, Board of Directors, Liaoning Province Chemistry and Chemical Engineering Society
栗万侯	LI Wan-hou	Member, Board of Directors, Liaoning Province Coal Society
邹德臣	TSOU Te-ch'en	Staff, Liaoning Province Foreign Affairs Bureau
邓小辉	TENG Hsiao-hui	Staff and Interpreter Liaoning Province Foreign Affairs Bureau

Fushun Banquet (May 29) Hosts and Participants:

李大章	LI Ta-chang	Chairman, Fushun Science and Technology Committee (Chief Host)
林耀森	LIN Yao-sen	Deputy Chief, Fushun City Foreign Affairs Bureau
罗荣本	LO Jung-pen	Staff, Fushun City Foreign Affairs Bureau
杜承印	TU Ch'eng-yin	Staff, Fushun City Foreign Affairs Bureau

24. Refinery Complex No. 2, Fushun—May 29, 1978

陈永寿	CH'EN Yung-shou	Deputy Refinery Manager
李洪才	LI Hung-ts'ai	Catalysis Workshop Technician
代承远	TAI Ch'eng-yuan	Engineer
尹炳	YIN Ping	Platinum Reforming Workshop Engineer
丁中	TING Chung	Engineer
杨瑾	YANG Chin	Engineer
王同力	WANG T'ung-li	Office Director
崔朝云	TS'UI Chao-yun	Staff, Refinery Office

25. West Open Pit Coal Mine, Fushun

高国祥	KAO Kuo-hsiang	Chief Engineer
罗长福	LO Ch'ang-fu	Engineer
李文山	LI Wen-shan	Engineer

Shanghai—May 31–June 5, 1978

王应睐	*WANG Ying-lai	Director, Shanghai Branch, CAS (Banquet Host, June 2)
梁文骅	*LIANG Wen-hua	Responsible Person, Office of General Affairs, Shanghai Branch, CAS
梁国志	LIANG Kuo-chih	Staff, Office of Foreign Affairs, Shanghai Branch, CAS
刘泽蔚	*LIU Tse-wei	Head, Foreign Affairs Office, Shanghai Branch, CAS
江征帆	*CHIANG Cheng-fan	Responsible Person, Shanghai Science and Technology Association

* Attended banquet, June 2

26. Institute of Organic Chemistry, Shanghai—June 1, 1978

	WANG Yü	Director (not met)
黄维垣	*HUANG Wei-yuan	Professor and Deputy Director
黄耀曾	HUANG Yao-tseng	Professor and Deputy Director Appointee
蒋锡夔	*CHIANG Hsi-k'ui	Professor
戴立信	*TAI Li-hsin	Associate Professor and Scientific Secretary
夏宗芗	HSIA Tsung-hsiang (f)	Researcher
姚介兴	YAO Chieh-hsing	Chief, Scientific Administrative Group
王大胜	WANG Ta-sheng	Protein Synthesis Group Leader
屠传忠	T'U Ch'uan-chung	Plasma Substitute Laboratory
周凤仪	CHOU Feng-yi (f)	Synthesis of Polyribonucleotide Laboratory
陈海宝	CH'EN Hai-pao	Synthesis of Polyribonucleotide Laboratory
金善炜	CHIN Shan-wei	Trichosanthin Structure Determination Laboratory
吴照华	WU Chao-hua (f)	Structure and Reaction of Arteannium Laboratory
王永禄	WANG Yung-lu	CO_2 Laser Chromatography Laboratory
戴行义	TAI Hsing-yi	Polymer Physical Chemistry Laboratory
史观一	SHIH Kuan-yi	Stress Cracking Resistance Laboratory
吴厚铭	WU Hou-ming	Liquid Crystals Laboratory
陈庆云	CH'EN Ch'ing-yun	Synthesis and Applications of Oxa-perfluoralkyl Sulfonic Acids Laboratory
徐珍娥	*HSU Chen-o (f)	Staff and Interpreter, Foreign Affairs Office

* Attended banquet, June 2

27. Institute of Biochemistry, Shanghai

王应睐	*WANG Ying-lai	Director

28. Institute of Nuclear Research, Shanghai—June 1, 1978

CHIN Ho-tsu	Director (not met)
SHIH Hsuan-wei	Deputy Director and Physicist (not met)
CHAO Chung-hsin	Chief Engineer and Radiochemist
CHANG Chia-hua	Professor and Responsible Person
LIU Nien-yen	Professor, Radiation Chemistry
CHU Chin-yi	Radiation Chemistry
LI Yun-che	Nuclear Chemistry
SHA Chen-yuan	Staff Officer
LIU Ken-pao	Nuclear Physics, Research Assistant
CH'EN Mo-pai	Engineer
CH'ANG Hung-chun	Cyclotron Director
NI Hsing-po	Perturbed Angular Correlation Laboratory
LI Ming-ch'en	Nuclear Physicist
LING Shen-hao	Nuclear Chemist
YAO Fu-chun	Section Head, Tritium Labeling of Amino Acids
YEN Yi-fa (f)	
TENG Liang-ch'ing	Cobalt-60 Source
LING Ching-hsin	Nuclear Detectors

29. Institute of Materia Medica, Shanghai—June 1, 1978

PAI Tung-lu

* Attended banquet, June 2

30. Futan University, Shanghai—June 2, 1978

	SU Pu-ch'in	President (not met)
郑子文	CHENG Tzu-wen	Vice President
蔡传廉	TS'AI Ch'uan-lien	Chief, President's Office and Scientific Research Bureau
郑绍濂	CHENG Shao-lien	Responsible Person, Department of Scientific Research Affairs
卢鹤绂	*LU Ho-fu	Professor of Physics
吴浩青	WU Hao-ch'ing	Professor of Electrochemistry; Associate Chairman, Department of Chemistry
顾翼东	KU Yi-tung	Professor of Inorganic Chemistry
于同隐	*YÜ T'ung-yin	Professor of Polymer Chemistry
陈文涵	CH'EN Wen-han	Associate Chairman, Department of Chemistry
邓家祺	TENG Chia-ch'i	Associate Professor of Analytical Chemistry
高滋	KAO Tsu (f)	Associate Professor of Catalysis
费伦	FEI Lun	Lecturer in Physical Chemistry
秦启宗	CH'IN Ch'i-tsung	Associate Professor of Nuclear Chemistry
蔡祖泉	TS'AI Tsu-ch'üan	Responsible Person, Department of Optics
陈星华	CH'EN Hsing-hua	Department of Optics
李富铭	LI Fu-ming	Department of Optics, Laser Division
徐凌云	HSU Ling-yun	Briefer, SBS Thermoplastics Elastomer Laboratory
何曼君	HO Man-chün	SBS Thermoplastics Elastomer Laboratory
朱文炫	CHU Wen-hsuan	ABS Copolymer Laboratory

* Attended banquet, June 2

张中权	CHANG Chung-ch'üan	ABS Copolymer Laboratory
王立惠	WANG Li-hui	Thermal Analysis Laboratory

Hung Ch'iao People's Commune—June 2, 1978
TAO Shao-pao

31. Shanghai Institute of Chemical Engineering, Shanghai—June 1–2, 1978

*WANG Ch'eng-ming	Professor (working in area of corrosion)
WU Hsing-chiu	(working in area of corrosion)
LI Shih-chin	Professor and Polymer Program Head
CHAO Te-jen	Associate Professor, Polymer Program
WU Ho-jung	Lecturer, Polymer Program
LOU K.-t.	Lecturer, Polymer Program

32. General Petrochemical Works, Shanghai—June 5, 1978

HSI Yi-chun	General Engineer
LENG Shao-yung	Chief, Technical Department
WU Chia-chi	Technical Department
HUANG Wei-ming	Chief, Contract Division, Foreign Affairs Office
TS'AO Wei-lien	Atmospheric Distillation Unit
TS'AO Shih-wu	Atmospheric Distillation Unit
WANG Tsung-yi	Engineer
CHANG Fu-hsing	Engineer and Chief, Acetaldehyde Workshop

* Attended banquet, June 2

33. Institute of Ceramic Chemistry and Technology, Shanghai—June 5, 1978

YEN Tung-sheng	Director and Vice President, Shanghai Branch, CAS
YÜ Shen-ching	
YING Chih-wen	
CH'I Chin-chih	
WEN Shu-lin	
CHU Ping-ho	Ferro-electric Materials
TING Hui-li	Electrically Induced Birefringence of Ferro-Electric Materials
CHIN Chi-jen	Production of Lithium Niobate
HSUEH Wen-lung	Silicon Nitrides
T'AN Hao-jan	Laser Apparatus
WU Chung-jen	Noncrystalline Amorphous Semiconducting Materials
KUO Ch'ang-lin	X-Ray Diffraction Apparatus
LI Chia-shih	Porcelain Preservation
CH'EN Hsien-chiu	Porcelain Preservation

34. Shanghai Normal University, Shanghai (discussion on chemistry education)—June 5, 1978

Participants from the Normal University:

夏炎	HSIA Yen	Professor of Chemistry and Deputy Chairman, Academic Affairs Committee of the University
顾可权	KU K'o-ch'üan	Professor of Organic Chemistry
潘道皑	P'AN Tao-ch'ien	Associate Professor of Chemistry
周乃扶	CHOU Nai-fu	Associate Professor of Chemistry
吴欣然	WU Hsing-jan	Lecturer, Department of Chemistry

杨维达	YANG Wei-ta	Lecturer, Department of Chemistry
范杰	FAN Chieh	Teaching Assistant, Department of Chemistry
陈良	CH'EN Liang	Teaching Assistant, Department of Chemistry
徐伯隆	HSU Po-lung	Teaching Assistant, Department of Chemistry
朱兵	CHU Ping	Chemistry Department

Participants from the Shanghai Institute of Chemical Engineering:

李国镇	LI Kuo-chen	Associate Professor
路琼华	LU Ching-hua	Associate Professor
程镇达	CH'ENG Chen-ta	Teaching Assistant

Participants from Futan University (all Department of Chemistry):

吴浩清	WU Hao-ch'ing	Professor
邓景发	TENG Ching-fa	Associate Professor
徐和功	HSU Ho-kung	Associate Professor
陶增宇	T'AO Tseng-yü	Lecturer
谢高阳	HSIEH Kao-yang	Lecturer
金若水	CHIN Jo-shui	Teaching Assistant
郑成法	CHENG Ch'eng-fa	Teaching Assistant
李启东	LI Ch'i-tung	Teaching Assistant
叶明吕	YEH Ming-lü	Teaching Assistant
朱伯卿	CHU Po-ch'ing	Teaching Assistant

Participants from the Shanghai Teachers' Institute (two-year school):

吴迪胜	WU Ti-sheng	Lecturer
邱先新	CH'IU Hsien-hsin	Teaching Assistant
曹锦荣	TS'AO Chin-jung	Teaching Assistant

35. Shanghai Water Works—June 5, 1978

SUNG Kuang-yuan	Deputy Manager

36. Computer Technology Research Institute, CAS, Shanghai Branch—June 5, 1978

LI Tzu-ts'ai	Director, Eighth Section
WANG Chia-te	Working in Chemistry Field
LI Ch'ing-chih (f)	Working in Chemistry Field
CHIANG Nai-hsiung	

Hangchow—June 3–4, 1978

王杰	*WANG Chieh	Deputy Director, Chekiang Province Science and Technology Committee (Banquet Host, June 3)
张怀礼	*CHANG Huai-li	Administrative Office Chief, Chekiang Province Science and Technology Committee
马安东	*MA An-tung	Office Secretary, Chekiang Province Science and Technology Committee

39. Chekiang University, Department of Chemical Engineering—June 4, 1978

杨士林	*YANG Shih-lin	Professor
周春晖	*CHOU Ch'un-hui	Professor

* Attended banquet, June 3

Sian—June 6–7, 1978

杨戈	*YANG Kao	Chairman, Science and Technology Committee, Shensi Province (Banquet Host)
秦克甲	*CH'IN K'o-chia	Chief, Administrative Office, Science and Technology Committee, Shensi Province
韩锐民	*HAN Jui-min	Staff Administrative Office, Science and Technology Committee, Shensi Province
杨明	*YANG Ming	Administrative Office, Science and Technology Committee, Shensi Province
程燕如	*CH'ENG Yen-ju	Foreign Affairs Bureau, Shensi Province

40. Northwest University, Sian—June 7, 1978

刘健	LIU Chien	Vice-President
張伯声	CHANG Po-sheng	Vice-President and Professor, Department of Geography
王铁民	WANG T'ieh-min	Deputy Chief, Administrative Office
刘顺康	LIU Shun-k'ang	Deputy Chief, Scientific Research Division
陈运生	*CH'EN Yun-sheng	Associate Professor and Deputy Chairman, Department of Chemistry
刘衎烈	*LIU Kan-lieh	Associate Professor and Deputy Chairman, Department of Chemical Engineering
李轼	LI Shih	Associate Professor, Department of Chemistry

* Attended banquet, June 6

李铸	LI Chu	Associate Professor, Department of Chemistry
孙聚昌	SUN Chü-ch'ang	Associate Professor, Department of Chemistry
胡荫华	HU Yin-hua (f)	Lecturer, Department of Chemistry
刘源发	LIU Yuan-fa	Lecturer, Department of Chemistry
马宝歧	MA Pao-ch'i	Teaching Assistant, Department of Chemistry
祖庸	TSU Yung	Teaching Assistant, Department of Chemistry
王善学	WANG Shan-hsueh	Teaching Assistant, Department of Chemistry
	YANG Mao-hsing	Librarian
	TU Wen-hu	Atomic Absorption Spectroscopy
	SUNG Ti-hung	Head, Thermochemistry Laboratory

Lanchow—June 8–10, 1978

	*CHANG Chin-yi	Deputy Director, Science and Technology Committee, Kansu Province (Banquet Host, June 8)
尹仲礼	*YIN Chung-li	Chief, Office of General Affairs, Science and Technology Committee, Kansu Province
刘三明	*LIU San-ming	Staff, Science and Technology Committee, Kansu Province
刘继顺	*LIU Chi-shun	Staff, Foreign Affairs Bureau, Kansu Provincial Government
	CHIAO Chien-min	Staff, Foreign Affairs Bureau, Kansu Provincial Government

* Attended banquet, June 8

41. Institute of Chemical Physics, Lanchow—June 8, 1978

申松昌	*SHEN Sung-ch'ang	Deputy Director
李树维	LI Shu-wei	Director, Research Division
尹元根	*YIN Yuan-ken	Chief, No. 4 Research Group (oxidative dehydrogenation catalysis, temperature programmed desorption, catalytic kinetics)
丁时鑫	TING Shih-hsin	Group Leader, Oxidative Dehydrogenation Catalysis
徐慧珍	HSU Hui-chen (f)	Group Leader, Temperature Programmed Desorption
丁雪茄	TING Hsueh-chia (f)	Group Leader, Catalytic Kinetics
陈英武	CH'EN Ying-wu	Chief, No. 5 Research Group (oxosynthesis catalyst preparation)
杨振宇	YANG Chen-yü	Group Leader (oxosynthesis catalyst preparation)
俞惟乐	YÜ Wei-lo (f)	Chief, No. 1 Research Group (GC glass capillary columns, GC/MS computer systems)
吕祖芳	LÜ Tsu-fang (f)	Group Leader, GC Glass Capillary Columns
顾文华	KU Wen-hua	Group Leader, GC/MS Computer System
金道森	CHIN Tao-sen	Chief, No. 3 Research Group (crown and cryptate compounds)
夏远椒	HSIA Yuan-chiao	Group Leader (crown and cryptate compounds)
姚钟麒	YAO Chung-ch'i	Group Leader (crown and cryptate compounds)

* Attended banquet, June 8

潘华山	P'AN Hua-shan	Director, Scientific Information Division
江天籁	CHIANG T'ien-lai (f)	Interpreter

42. Institute of Modern Physics, Lanchow—June 8, 1978

杨澄中	*YANG Ch'eng-chung	Director
许士元	HSU Shih-yuan	Briefer, Cyclotron
李学宽	LI Hsueh-k'uan	Cyclotron Operator
	LI Chen	Cyclotron Operator
朱学正	CHU Hsueh-cheng	Cyclotron Operator
乌恩九	WU En-chiu	Section Chief, Nuclear Physics
代光牺	TAI Kuang-hsi	Nuclear Physics Group (also Interpreter)
孙锡军	SUN Hsi-chün	Nuclear Physics Group
郭俊盛	KUO Chün-sheng	Nuclear Physics Group
诸永泰	CHU Yung-t'ai	Nuclear Physicist, Elastic Scattering (carbon on carbon), Accelerator Development (also Interpreter)
范国英	FAN Kuo-ying	Elastic Scattering
吴中立	WU Chung-li	Elastic Scattering
关铎	KUAN To	Accelerator Development (Section Chief)
张恩厚	CHANG En-hou	Accelerator Development
未宝文	WEI Pao-wen	Accelerator Development
叶维祎	YEH Wei-yi	Accelerator Development
王树芳	WANG Shu-fang (f)	Section Chief, Radiochemistry
李文新	LI Wen-hsin	Briefer, Radiochemistry
范我	Fan Wo	Briefer, Radiochemistry
周志明	CHOU Chih-ming	Briefer, Radiochemistry Applications
马希亮	MA Hsi-liang	Computer Center
陈安义	CH'EN An-yi	Computer Center
	HUI Ning	Heavy Ion Radiochemistry Group

* Attended banquet, June 8

	SUN Tung-yu	Heavy Ion Radiochemistry Group
	SUN Hsu-hua	Heavy Ion Radiochemistry Group
张锦添	CHANG Chin-t'ien	Detectors
张敏	CHANG Min (f)	Staff
	LIU Chia-wen (f)	Secretary

Lanchow University (not visited)

	HSU Kung-o	Professor and Theoretical Nuclear Physicist

43. Academy of the Lanchow Chemical Industry, Lanchow—June 8, 1978

邓大华	TENG Ta-hua	Director
陈信华	*CH'EN Hsin-hua	Deputy Director and Engineer
	SO Yi-lin	Vice-Director
	SU Tien-kung	Vice-Director
	MU Shu-jen	
	WU Ti-hua	
	CHOU Yü-lan (f)	Total Organic Analysis of Liquids Using Infrared
	CHAO Chun (f)	Anaerobic Degestion
	NIEN Shu-k'ai	Corrosion
	LIN Hsiao-kuang	Electrochemistry
	CHANG Shu-chen	
	HO Yi	X-ray Diffraction
	HU Ch'un-mei (f)	Transmission Electron Microscopy

44. Lanchow Institute of Petroleum Research, Lanchow—June 8, 1978

余志英	*YÜ Chih-ying	Director

46. Liuchiahsia Hydroelectric Plant—June 10, 1978

李文忠	LI Wen-chung	Chief Engineer
卫克俭	WEI K'o-chien	Vice Chairman, Revolutionary Committee of the Plant

* Attended banquet, June 8

APPENDIX D: Description of Programs at Individual Institutions

1. Institute of Chemistry, Peking

The Peking Institute of Chemistry of the Academy of Sciences was started in 1956 with emphasis on studies of organic, physical, polymer, inorganic, and analytical chemistry. Its work has led to the formation of two other institutes of the Academy—the Institute of Environmental Chemistry and the Institute of Photography. Of its 700 staff, 300 are scientists with 40 having a Ph.D. degree or its equivalent.

The instrumentation of note at this institute includes a Hilger X-ray generator with a Nonius Weissenberg camera. There is also an AEI mass spectrometer including a PDP-8E computer plus disc. Other instrumentation available includes a He–Ne laser of West German manufacture for holography, a vacuum deposition system of Chinese manufacture, gel permeation chromatography, HPLC (a good quality instrument of Chinese design), a Supercon NMR of French manufacture (RMN-250, Camaca, 1974), a Perkin-Elmer 180 IR equipped for polarization measurements, an "Instron" type device, a dynamic mechanical spectrometer, a tortion braid apparatus of Chinese manufacture, and a thermogravimetric analyzer of Japanese manufacture. A home-built ESR is also reported to be available at this institute. The RMN-250 Supercon NMR is apparently not currently operational because of difficulties in providing servicing.

2. Institute of Physics, Peking

The Institute of Physics in Peking was set up in 1950, at which time the total staff was 45. The present staff in all categories is 1200 people. The institute conducts research in semiconductor physics, infrared physics, metallurgy, plasma physics, solid-state physics, low-temperature physics, high-pressure physics, lasers, acoustics, and theoretical physics.

A helium liquefier is available with a capacity of 20 liters per hour. The liquefier was built in the institute. It utilizes an expansion engine of the Collins type. They have a superconducting magnet that can produce fields up to 110,000 Gauss, as well as a ^3He–^4He dilution refrigerator capable of producing temperatures down to 0.04°K. The compressor for their helium liquefier was made in Lanchow, while their liquid helium Dewars are imported from France.

The principal areas of work in theoretical physics involve gravitational waves and elementary particles. At this laboratory we learned of the plans to build a high-energy proton accelerator in Peking (not at the Institute of Physics) that would duplicate the Brookhaven AGS. This is relevant to Item 7 in the program plan that was evolved at the National Science Conference for the year 2000.

3. Peking University, Peking

Peking University was founded in 1898 and celebrated its 80th anniversary two weeks before our visit. It was the cradle of the May 4th movement, representing the beginning of the democratic revolution in China. The university has developed enormously since its founding and now has 22 departments with 2800 staff and 6400 students. In 1978, 1800 undergraduates and 350 graduate students will enter the University. From 1949–66 there were 20,000 graduates of the university, many of whom are now in responsible positions.

During the Cultural Revolution, the undergraduate curriculum was reduced to three years and the graduate program was eliminated. In 1978–79 for the first time admission to a four-year program is by (national) entrance examination and graduate studies have been restored. While there is no Ph.D. degree, a three-year program involving a 1½-year thesis and leading to a diploma is planned. It may also be of interest to note that elementary education was reduced from 12 years to 10 (5 years of elementary school, 5 years of high school) during the Cultural Revolution. Students may now start university training prior to finishing high school if they can pass university entrance examinations. Entrance examinations for graduate studies are now given by individual universities; students can take several and indicate a choice of schools. The requirement that students must pass entrance examinations this year is expected to increase the general level of student quality. Entrance by students from families of manual workers is not now specially favored, but there are quotas for Chinese minority groups.

With the political demise of the Gang of Four, more emphasis is being placed on fundamental courses and research. Two-thirds of the university students are in the natural sciences. There is no engineering, medical, or agricultural school. The titles of Professor and Associate Professor were eliminated during the Cultural Revolution but have now been restored. Professors have tenure and often cannot leave their university position, even if they want to. Professors provide advice to industry and government but do not receive consulting fees. M.S. and Ph.D. degrees have been eliminated but diplomas are given upon completion of graduate studies.

The Department of Chemistry consists of 10 sections with 250 teachers (12 professors and associate professors) with 550 students. The chemistry faculty includes Professor Huang Tzu-ch'ing, a physical chemist who studied at MIT under Beattie, and Professor Tang Yu-chi, an X-ray crystallographer who studied at Cal Tech. First-year undergraduate students take courses in general chemistry, mathematics, physics, English, and politics; second-year students take analytical and organic chemistry, mathematics, physics, and English; in the third year they take organic, physical, structural and industrial chemistry; while fourth-year chemistry students do research and elect courses among those in advanced organic chemistry, advanced physical chemistry, polymer chemistry, colloid chemistry, and advanced inorganic chemistry. All students must spend one semester on a research project. Graduate students may come from Peking or other universities or from industry. Because of the large staff, graduate students do not serve as teaching assistants.

Some of the instrumentation available for support of research at this university includes X-ray powder cameras, a mass spectrometer (Varian-MAT, 1965) for N^{14}/N^{15} studies, a UNICAM IR (1964), and thermal diffusion columns.

4. Tsinghua University, Peking

Tsinghua started in 1911 as a high school that prepared students to study abroad. In 1925 it was changed to a university of social studies, liberal arts, and science. Then in 1952 it was changed to a university of technology. It has nine departments including electronics, mechanics, chemical engineering, construction, engineering physics, hydraulics, and architecture. It currently has a staff of 2800 and a student body of 7000. It also has 100 foreign students from 25 nations.

A Petroleum Refining Institute was established in Peking in 1952 and moved to Tsinghua in 1958. The Chemical Engineering department was also established in 1958. At that time nuclear chemical engineering and a high-polymer program were also established.

From 1961 to 1966 there was an average of 150 graduates per year. In the period of the Cultural Revolution from 1966 to 1969, no new students were admitted. From 1970 on new students were admitted, but with the disturbance from the Gang of Four no real educational program was possible, and the school shifted to a middle-school level practical program. Entrance exams have recently been reinstituted and the curriculum fixed at four years.

In chemical engineering there are sections in the areas of transport processes, reaction engineering, processing systems, high polymers, nuclear chemical engineering, and nonmetallic inorganic materials technology; they also plan to build a section of physical chemistry. Currently they have 170 freshmen and a total student body of 800. They plan to admit 15 graduate students in the fall of 1978. At present they have two professors, five associate professors, and 67 lecturers. The Department Head, Professor Wang Chia-ting studied under W. K. Lewis at MIT. The laboratories were said to be destroyed during the period when the Gang of Four was in power, and we did not see them. We were unable to obtain a copy of their current curriculum.

Some instrumentation at this university includes a SEM (JEOL, JSM-113), a mass spectrometer (Varian MAT M-86, gas analyzer), and a unit operations laboratory including specialized equipment for studying fluidized beds.

5. Institute of Atomic Energy, Peking

The Institute of Atomic Energy was founded in 1956. The current director is Wang Kan-chang. Ch'ien San-ch'iang, the long-time former director is now Vice-President of the Chinese Academy of Sciences. Ch'ien received his degree in nuclear physics from the Joliot Laboratory in Paris in the 1930s. During our visit we met two deputy directors: Li Shen-nan (we presently do not have any information on his background) and Wang Te-hsi. Wang was educated in chemical engineering at MIT and got his degree in about 1944 with W. K. Lewis. He is an expert on isotope separation science and technology.

There are approximately 1500 total personnel at the institute of which 600 are university graduates. The chemistry section has a total of about 350 people of whom approximately 175 are university graduates. The general programs of the institute are the following: reactor material testing, fuel reprocessing, theoretical nuclear physics, neutron physics, work with two accelerators, an electronics laboratory, theoretical heavy-ion physics, measurement of neutron cross-sections, and inelastic neutron scattering. The major facilities available at the institute are a 1.2-meter diameter cyclotron for the acceleration of protons and deuterons, a 600-kilovolt Cockroft–Walton D-T accelerator to produce 14 MeV neutrons, a 7-mW heavy-water reactor containing

2% uranium-235, built by the U.S.S.R., and a light-water swimming pool reactor containing 3% uranium-235. Construction of the swimming pool reactor started in 1960, and it was put in operation in 1964. The institute hopes to purchase a 20 MeV tandem accelerator for protons. It does not claim that any of its experimental facilities are at the forefront of modern nuclear science.

Some of the other instrumentation available to support the research work at this Institute includes an 800-channel multichannel analyzer imported from France, NaI and Au–Si detectors as well as some Ge–Li detectors, Calutron-electromagnetic isotope separators, remote control manipulators in hot cells, a 2.5-MeV Van de Graaf accelerator, a proton-induced X-ray fluorescence analysis system, and instrumentation for neutron activation analyses and charged-particle activation analysis.

6. Institute of Materia Medica, Peking

This institute was established in 1958 and currently has more than 700 staff, of whom 60% are research staff and technicians. The areas involved in the program of the institute include synthetic organic chemistry, phytochemistry, antibiotics, analytical chemistry, pharmacology, medicinal plants, and the cultivation of medicinal plants. The facilities include three pilot plants for the production of significant amounts of materials. There are approximately 40–50 synthetic organic chemists and 40 organic chemists working in phytochemistry. The detailed programs underway at this Institute are described in Chapter 5, Organic Chemistry.

Some of the instrumentation available to support the research work at the Institute includes a JEOL 60 MHz NMR (H^1, No C^{13}), an ancient single focusing MS, routine IR and GC, and a CS-900 dual wave TLC scanner of Chinese manufacture.

7. Institute of Biophysics, Peking

The Institute of Biophysics was established in 1958 and thus is a relatively new institute of the Academy of Sciences. It currently has a total complement of 400 workers. The principal areas of research include radiobiology using a cobalt-60 source. Research in this area includes long-term low dosage studies with monkeys, high dose short-term exposure studies, with both internal and external dosages. A second area of research includes biology and biophysics of receptors, including visual receptors and vibrational receptors in the legs of pigeons. A third area of research is molecular biology, including

(a) X-ray diffraction studies of proteins (including the refinement of the structure of insulin to 1.8 Å, plus insulin analogs),
(b) nucleic acids (including the chemical and enzymatic synthesis of polynucleic acids and *t*-RNA synthesis),
(c) enzymes (with a focus on allostearic properties of enzymes such as glyceraldehyde phosphate dehydrogenase), and
(d) enzymatic diagnosis of cardiac disorders using creatine kinease.

The emphasis of the work in this institute is on pure research, but there are also some practical applications. The director of the molecular biology activities in this Institute is Cambridge-educated Tsou Chen-lu.

In addition to a Phillips four-circle X-ray diffractometer, there is also a JEM-7 electron microscope from JEOL, vintage 1965. This is a good microscope that appears to be in excellent condition; all service and maintenance are done by the Chinese. One of the interesting parts of our tour included the shop facilities of the institute. The shops appear to be active and well staffed; we were shown several machine shops and a very busy glass blowing shop.

8. Petrochemical Research Institute, Peking

This Institute was established in 1958. They currently have a research staff of 800 with 60% college graduates. The Deputy Director, Dr. Wu Pao-chen, received his Sc.D. from Hottel and Williams at MIT. In order to meet the requirements of the petroleum industry, the institute runs a night school for engineers, college graduates, and workers. This school teaches English to students and physical chemistry plus higher mathematics to college graduates. In order to obtain a high caliber staff, they train graduate students who take basic courses at the nearby universities. They currently have 20 from all over the country. The institute works basically on the problems of the Peking Petrochemical Complex. When necessary, a task force team is established with the plant. The institute also initiates some work and at times "propagandizes" the plant to try out innovations launched at the institute. If they have a long-term fundamental problem, they attempt to interest a university professor. Such people occasionally lecture at the institute.

The institute appears to be reasonably well instrumented to carry out its mission, and they have among other instruments the following: BET for surface areas, pore-size distribution via a sorptomatic instrument (Carlo Erba, Italy), multiple GC units made by the Peking Analytical Co., mercury porosimeter (Carlo Erba, Italy), a SPEX laser Raman, an atomic absorption spectrophotometer (made in China), and a capillary column GC (made by Peking Analytical Co.).

9. Peking General Petrochemical Works

Construction of the Peking Petrochemical Complex began in 1968. It currently consists of a refinery and six chemical plants. There is also a chemical engineering design institute (1200 engineers on site) and a research institute (in Peking) connected with the plant and reporting at least in part to the plant manager. It had an originally designed capacity of 2,500,000 ton/year but this has now been expanded to 7,500,000 ton/year. Feedstock is essentially Tach'ing crude, which is transported from the Tach'ing Oil Fields via a 1600-km pipeline. They also run a small and unknown quantity of Yumen crude. They have a total staff of some 11,800 with 37,000 people living in the local workers' villages. During construction they researched, designed, and built as much as they could; where this was not possible they imported the technology but attempted jointly to manage construction and use as much Chinese machinery as practicable. The plant also constructed hospitals, kindergartens, workers' apartments, and shopping facilities.

Refinery. Crude is first distilled in an atmospheric distillation unit and then vacuum distilled. The heavy gas oil is catalytically cracked in a fluid bed cracker using a zeolite catalyst and there is also a fixed bed reforming unit using a monometallic Pt catalyst, which operates on distillate from the atmospheric vacuum still. The petroleum refinery produces gasoline and diesel fuel, various lubricants from a lube plant, and a BTX cut from the reformer.

Chemical Plant. They have a gas oil cracker that produces ethylene, propylene, and aromatics at one of the chemical plant complexes and they produce 60–90 separate chemical products. These cover a broad spectrum and some are listed in Table VI. While much of the petroleum refining technology was Chinese-designed and built, the bulk of the chemical plants are licensed recent foreign technology, as noted in the table. They claim to have a total of 39 major processing units; of these, 7 are "imported."

TABLE VI
Peking Petrochemical Complex

Crude Source: Mainly Tach'ing
Supply Route: 1600-km heated pipeline
Crude Run: 7,000,000 tons/year
Workers: 11,800

Process	Capacity (t/yr)	Technology
Petroleum Refining		
Fluid bed cat. cracking	1,200,000	Chinese—zeolite
Fixed bed reforming	800,000	Chinese—Pt/Al_2O_3
Heavy fuel oil	200,000	Chinese—distillation
Gasoline	>1,000,000	
Diesel	2,000,000	
LPG	80,000	
Chemicals		
Ethylene	300,000	Lummus heaters
Polyethylene	180,000	Sumitomo—high pressure, low density
Polypropylene	80,000	Chinese—$TiCl_3$
Ammonia	15,000	Kellogg
cis-Polybutadiene	60,000	Chinese
Polystyrene	1,200	
Toluene	2–5,000	
p-Xylene	45,000	UOP
Phenol/Acetone	10,000	
Butanol	8,000	
2-Ethylhexanol	10,000	Japanese
Alkylbenzene	7,500	
NH_4NO_3	22,000	
Nitric acid	35,000	
Ethylene glycol	40,000	Foreign
Power station	Two producing	45,000 kW
Waste water treating plant	Three	

10. Institute of Environmental Chemistry, Peking

The Institute of Environmental Chemistry was formed in March 1975, in part with personnel from the Institute of Chemistry. Its purpose is to concern itself with pollution and environmental technology problems in China. The institute, with only 200 employees, is still in a formative stage. However, they have a significant amount of new instrumentation (atomic absorption, VPC, IR emission spectrometer, X-ray fluorescence, inorganic chromatography, etc.). Most of the instruments are provided by the manufacturing industries in return for research and services.

11. Institute of Photography, Peking

Although no one from the delegation visited this institute, it was frequently referred to as an institution where high quality research in chemistry was being carried out. It is apparently organized into four divisions, two dealing with the chemistry of photography, and two divisions dealing with photochemistry. The Institute of Photography has the only Fourier transform NMR spectrometer with ^{13}C capability in China (a Varian XL-100).

N. C. Yang from the Department of Chemistry, University of Chicago, spent several months visiting this institute in 1978.

12. Peking Vinylon Plant

A visit was made to a vinylon (polyvinyl alcohol (PVA)) fiber factory in the vicinity of Peking. Vinylon is much more important in China (as in Japan), where it is used as a cotton substitute, than it is in the United States. Cotton is still the most widely used fiber; but because its wide usage places a heavy demand on agricultural resources, efforts are being made to substitute synthetics such as PVA and polypropylene. Nylon-6 polyethylene terephthalate and acrylic are being produced but are relatively expensive for wide use. Nylon-66 production is being initiated using Japanese or German technology. Cotton use is estimated to be 2–3 times that of synthetics, and PVA, PET, and PAN consumption are similar but nylon-6 is less. Polyester–cotton blends are beginning to be used for better quality clothing items (shirts, "Mao-jackets"), but these blends cost 1.5 to 2 times more than cotton. Polyester use is increasing appreciably. Viscose rayon is in use for clothing, but its use is decreasing. Rayon is the more widely used fiber for tire cords, but polyester and nylon cord use is growing.

Use of steel cords is negligible. Aromatic polyamides are under study, but there is no commercial production.

The vinylon factory was built in 1963 utilizing Japanese technology. It was designed with a capacity of 11,000 metric tons per year, but with Chinese improvements the capacity was increased to 18,000 tons. The plant represents an investment of Y 89,000,000, occupies 460,000 square meters, and employs 2150 workers. It produces two kinds of products, PVA staple and "tow" or long fiber. The former is primarily for clothing applications (mixed with cotton), while the latter is used for industrial purposes such as tire cord, machinery belting and fishnet. The factory buys its PVA (DP = 1700) from a Peking supplier who prepares it by hydrolysis of polyvinyl acetate. This is prepared from acetylene, which is derived from calcium carbide, which in turn is produced from coal. It is washed to remove sodium acetate, dissolved in hot water (98°C) to form a 16% solution, filtered, degassed, and then wet spun through a 6000-hole spinerette into a sodium sulfate coagulating bath. It is drawn and heat treated in a wet and dry step, chemically treated with formaldehyde and sulfuric acid to reduce solubility and improved heat resistance, then dried and packed.

The factory seems to be reasonably well mechanized and is not overly labor intensive. Among the workers are 150 university trained people, the rest being middle school graduates (17–20 years of age) who were trained at the factory. The average age of the workers is 30.

No appreciable research is done at the factory. Quality control is modest, involving viscosity measurements on PVA solutions and tensile strength measurements (using Japanese instruments). It was pointed out that nine new PVA plants are being set up, the more recent of which are of Chinese design. This plant is quite comparable to many similar small, low-technology polymer processing operations in the United States.

13. Institute of Chemical Physics, Talien

The Institute of Chemical Physics in Talien was set up by the Japanese in 1908 as a technical institute. In 1950 after Liberation, it became the Petroleum Research and Engineering Institute of the Chinese Academy of Sciences. To meet the industrial expansion goals of the time, the institute set up research laboratories for chemical engineering, catalysis, synthetic liquid fuels, and modern analytical chemistry. After the initial reconstruction phase in China, the Chinese Academy of Sciences changed it into the Institute of Chemical Physics in 1962. The Director is Professor Fan Ta-yin. At the present time there are

800 people working in the Institute; 500 are scientists, 200 supporting staff, and 100 other administrative personnel.

There are nine research laboratories in this institute. One is the basic and applied catalysis laboratory, where they are doing research on the physical chemical properties of catalysts and studies of the adsorbed state. The catalysis program is primarily directed at petroleum reformation and nitrogen fixation. The second laboratory is concerned with chromatography and general analytical chemistry. The work on gas chromatography involves high-efficiency columns as well as the theory of gas chromatography. In the third laboratory they are working on molecular collision dynamics. They are just preparing to launch this project and plan to work on energy transfer processes and the theoretical analysis of scattering dynamics. The fourth laboratory is chemical engineering. They will be investigating reverse osmosis (for desalination) and chemical reaction engineering. In the fifth laboratory they are working on scientific instrumentation, computerization, and automation.

These five laboratories are located in the main building; there are four additional divisions in a laboratory unit near the beach. These include programs for the chemical simulation of nitrogen fixation; the study of olefin polymerization; the development of new reactions of transition metal catalysts; and the study of adsorbed states, trace analysis of organic materials, high pressure chromatography, and ultra analysis.

The library, started by the Japanese, currently subscribes to approximately 405 periodicals, 250 of which come from abroad and about 50 of which are of Russian origin. Western periodicals arrive on the order of three months after publication. Their holdings of *JACS* are continuous from the beginning. Their holdings of *Chemical Abstracts* start with 1907.

Some of the instruments available for research in this institute include GCs with flame ionization and electron capture detectors; a GC with an interface to an Atlas CUA MS; an electron microscope (JEOL) with 800×10^3 magnification and 70 Å resolution; a Perkin–Elmer 577 IR; an ESR spectrometer; and a DTS-130 computer (Chinese manufactured) with 16 K core, 16-bit words, basic processing time of 100 μsec, and a line printer.

14. Talien New Harbor

The members of our delegation were the first Americans to visit the new oil harbor at Talien. Its main function is to export petroleum. The harbor was constructed at the direction of Chou En-lai to meet the new requirements of China and was put into operation on May 1, 1976. All the equipment at the port was designed and built in China in a total period of 18 to 20 months. There is a bridge out to the docks that is 1.7 km long. They can load crude oil at a rate of 10,000 tons/hr, handle cargo boats of 100,000 tons, and have a storage depot for 300,000 tons of petroleum that comes down by pipelines from Tach'ing. This is the first oil port in China. At present they are exporting oil to Japan, to Romania via the Suez Canal, and to South Asia, and are interested in exporting petroleum to the United States. In the harbor we saw a tanker (see Fig. 25) of 100,000-ton capacity. Out in the bay there was another tanker of 50,000-ton capacity. The Chinese said they do not yet have the capability to load a tanker at sea.

15. Talien Locomotive Factory

A number of the members of the delegation visited the diesel locomotive factory in Talien. This plant was started by the Russians during their occupation of Manchuria in 1901. It was taken over by the Japanese and operated by them for the period from 1904 to 1945. After World War II the operation of the plant was taken over by the Chinese with Russian supervision. The plant was primarily a repair facility until 1954, when work was started to convert it to a production facility. First production started in 1956. By 1958 they were producing steam locomotives with a rating of 2000 horsepower.

After the Russians left in 1960, a group of Chinese engineers from the universities and institutes of the Academy of Sciences came to the locomotive works and designed a new production facility. By 1964 they had in production 2000-horsepower diesel locomotives of their own design. By 1969 they started the production of 4000-horsepower locomotives, which is their current model.

The plant for the production of 4000-horsepower locomotives has a variety of equipment. During the period from 1964 to 1978, they designed, revised, and built some 2000 production machines. They have approximately 60 automatic production lines for their production facilities. Although the work on the production of 4000-horsepower locomotives started in 1969, the first actual locomotive of this capacity did not roll off the assembly line until 1974. Much of this delay in start-up is attributed to poor labor–management relations during the period of the Gang of Four.

The locomotive works now engages 9000 workers of whom 2000 are women. There is a variety of machinery in the production facility. Some of the new large machines come from Japan. However, a large crankshaft turning lathe from the U.S.S.R. is still in use. Some recent additions to their production facilities include computer-controlled manufacturing machinery. The latter were completely designed and fabricated at the locomotive works. At present they manufacture everything required for the locomotives at that particular site, with the exception of bearings. Their annual production is 150 locomotives. The finished product is a very compact and well-finished unit. The locomotives are well designed from the point of view of ease of operation by the engineer in charge. They are very easy to start and the controls are very well laid out on the control panel.

16. No. 7 Oil Refinery, Talien

The Talien petroleum refinery has a long history. It is 44 years old and was originally planned and operated by the Japanese. The plant consists of two parts: the old plant, which has a capacity of 50,000 tons of crude oil per year; and a new part, which processes 5,000,000 tons of crude per year.

Until 1960 Russian oil was processed in the old plant, but when the relations were severed, the oil imported into Talien from the Soviet Union stopped. The discovery and the development of the Tach'ing Oil Field, which started in 1959–60 was the impetus for the modernization and enlargement of this refinery.

The present refinery has catalytic cracking units, reforming units, and distillation equipment. It produces a total of 1,200,000 tons of gasoline per year. The reforming plant has towers that are 30 meters tall and 2.6 meters in diameter. They were designed and built in China.

The new plant has 24 operating units. While the old plant is in poor condition, the new plant has excellent maintenance. The reforming section has an input of 600,000 tons of crude oil per year.

17. Tach'ing Academy of Science, Technology, and Design

The mission of this laboratory appears to be to provide technical service to the oil field production program at Tach'ing. Most studies are oriented toward the study of physical properties such as porosity and wetability of subsurface rock samples in contact with crude oil. The most interesting technique for measuring porosity was a 25-GHz microwave device for measuring dielectric loss from samples of rock formations exposed to water. We were also shown a centrifuge system for measuring the displacement of oil in porous samples by water at 21,000 rpm.

18. Tach'ing General Petrochemical Works

The Tach'ing Oil Field and Petrochemical Complex located in the northeast province of Heilungkiang is the showplace industrial complex in China. "Learn from Tach'ing" is an ever-present slogan. The first well was drilled in April 1959 and the field currently consists of several thousand wells. Since 1960 production has increased about 52-fold. All the wells observed were naturally flowing and were said to average between 20–100 tons/day of oil. Oil temperature averages 80°C and associated dry gas averages 3800 m^3/day (mainly CH_4). The one well observed in detail was stated to have an average production of 60–80 tons/day and this average production figure was steady over a 17-year production interval. Average well depth was stated at 3000 ft and the field was claimed to be about the size of the U.S. East Texas field. The Chinese expect to recover about 50% of the reserves and they are depleting the field at the rate of about 2.5%/year, yielding a 40-year field lifetime. Their cumulative production to date is about 25% of the current reserves.

During oil production some of the associated dry gas is separated from the crude and used to heat the oil. This heating is essential, since the crude has a high paraffin content and wax tends to precipitate unless the crude is maintained at elevated temperatures. Water, which is low in salt, is reinjected into the well to maintain oil field pressure.

The oil field consists of a series of oil pool layers interspersed between sandstone layers. Water is injected at different levels due to the differing oil viscosities at each level. The water-to-oil ratio is currently about 2:1. They drill about two wells/km^2.

The Tach'ing complex consists of the oil field, a petroleum refinery, and a chemical plant. The chemical plant is located about 20 km from the refinery. The reason given for the location spread was that there are plans to build other units in the intervening space. The complex includes housing, schools, hospitals, and shops. With a total population of 600,000 people, about 100,000 are stated to work in the oil field.

Petroleum Refinery. The petroleum refinery was constructed in stages, with the chemical plant being built at a secondary stage. From 1958–63 an atmospheric and vacuum distillation column and a thermal cracking unit were built. In 1963–64 a zeolite fluid bed catalytic cracking unit, a fixed bed platinum reforming unit, and a delayed coker were constructed. At this time the refinery had a capacity of 2,500,000 tons/year. This capacity has since been expanded to 5,000,000 tons/year. There is also a hydrogen plant, a cobalt/molybdenum hydrocracker operating on a mid-distillate. The resulting hydrofined product is used as reformer feed. There is a molecular sieve wax plant. The technology is essentially Western but designed and constructed by the Chinese. They are planning a lube oil factory and a steam cracker for ethylene production.

Chemical Complex. The chemical complex consists of a number of plants producing chemicals and fertilizers. These include ammonia, urea, polyacrylonitrile, ammonium nitrate, benzene via toluene dealkylation over molecular sieves, and mixed xylene separation (para by molecular sieve and ortho/meta by distillation). Most of the technology in the chemical plant is foreign, as noted in Table VII.

TABLE VII
Tach'ing General Petrochemical Works

Crude Source: Tach'ing
Crude Run: 5,000,000 tons/year
Supply Route: Heated Pipeline
Workers: 3000

Process	Capacity, tons/year	Technology
Petroleum Refining		
Atmospheric Distillation		
Vacuum Distillation		
Fluid-Bed Cat. Cracking	800,000	
Fixed-Bed Reforming		
Delayed Coking		
Hydrogen Plant		
Hydrocracking		Cobalt/Molybdenum Catalyst
Dewaxing		
Chemicals		
NH_4NO_3		
Acrylonitrile	5000	
Ammonia	300,000	Kellogg
Urea	480,000	Dutch
Nitric Acid		
Ethylene	Planning Stage	
Benzene via Toluene Dealkylation		
p-Xylene		Mol. Sieve Sep.
o/*m*-Xylene by Distillation		

19. Research Institute of Tach'ing General Petrochemical Works

The Tach'ing Research Institute is an intrinsic part of the Tach'ing complex. It has 111 staff members, 90 of whom are college graduates. Very few have advanced training beyond undergraduate level. The institute began operations in 1970 and during the past seven years has trained personnel for the needs of the Tach'ing complex as well as for its own needs. Projects are generated by the complex, at the Ministry in Peking, and at the Institute. There is some, but not much, research on new processes, and the bulk of the work is on current problems. There is contact with other institutes and there has, for example, been transfer of results from the Shanghai Institute of Petroleum Chemistry to the Tach'ing Research Institute. They do not carry out design work. This is evidently done at the Institute of Design of the Petroleum Ministry in Peking. Apparently when units are constructed, a construction team is formed, for example, between petroleum and design, and when the unit is finished, this team will move to another construction project. At present they work on all parts of the national plan set in Peking by the Ministry of Petroleum. They are now mainly concerned with chemistry problems but hope to work in the future on issues related to computer control.

The institute has among other equipment the following: BET, TDA (Shimadzu), X-ray diffraction (Chinese), and a microcatalytic reactor connected to GC on-line analyzers.

In the past seven years, the institute has completed projects on several items which have been commercialized at Tach'ing, including separation of xylenes by absorption, simulation of moving beds for xylene separation, and demethylation (thermal) of toluene to produce benzene. They do applied research on the purification of wax via hydrogenation over tungsten catalysts. Interestingly, the U.S. standards are used for food-grade wax, and the standard UV absorption method is also used. They also do screening studies for oil additive photostabilizers (polybutenes). This research involves a simple heated chamber containing a UV source. The 0.01%–0.05% concentration level is the additive target level. Work on zeolite synthesis is also conducted at the institute.

20. Institute of Petrochemistry, Harbin

The Harbin Petrochemistry Institute was originally established in 1952 as the Heilungkiang Petroleum Chemistry Institute. There is a staff of 400, seven research groups, two pilot plants, and one mechanical shop. It claims to have close ties with other research institutes.

Research is currently concentrated on petroleum chemistry and adhesives in the following areas:

Aromatics' manufacture and use
Catalytic oxidation of olefins
Research on the production of acetaldehyde trimer
Adhesive research
Physical–chemical analysis

As of May 1978 most of the journals in the library were from before late 1977. An exception was the *Journal of the American Chemical Society,* where the issue from February 15, 1978 was available. The library appeared to be very good for the size of the institute, having, of course, an extensive collection of Chinese journals, what appeared to be a wide collection of Russian journals, and a large number of English-language journals. A quick look indicated that most of the familiar journals were there as reproductions with the exception of *Inorganic Chemistry* and the *Journal of Organometallic Chemistry.* A very cursory look at the book and past journal collection indicated that this was a well-equipped library that probably fulfilled most of the needs of institute personnel.

Instruments of some interest include a differential thermoanalysis apparatus with a Japanese XY plotter made by Watanabe, a British thermal balance by Stanton, another DTA from Hungary, and a Chinese-made modern weighing balance that looked like a copy of a Mettler balance. The analytical emission spectroscopy unit was partially Chinese and partially German-made. An infrared spectrometer coupled to a catalyst absorption device was made by Unicam. The powder diffractometer used a pumped tube, and it was made by Hilger. There were also some Nonius specialized powder cameras. The gas chromatography unit was a GC5A Japanese model made by Shimatzu. There were also Chinese gas chromatography units.

This is a well-organized and independent, if somewhat isolated, laboratory. The institute management appears to know what the important problems are. They appear to have good communications with Tach'ing and perhaps with other institutes of petroleum chemistry. The institute seems to have generally good facilities. The staff appears to be cheerful, happy, and eager to work. Overall this looks like an effective laboratory.

21. Kirin University, Ch'angch'un

Kirin University started as a College of Administration in Harbin in 1946. In 1950 after Liberation it was moved to Ch'angch'un and became Northeast People's University, with departments in only humanities and literature. Natural science departments were added in 1952, and the name of this comprehensive university was changed to Kirin University in 1958.

At present there are 11 departments (mathematics, physics, chemistry, semiconductors, electronic computers, Chinese, history, economics, law, foreign languages, and philosophy), and a total of 31 specialties or disciplines under these departments (21 in science and 10 in humanities and literature). The number of students was 3400 when we visited in May 1978. The second- and third-year students are in a three-year system, while the first-year students are in the new four-year system. In the summer of 1978, 1500 new undergraduate students were expected to enroll. In addition, 250 graduate students or teachers in retraining programs will be admitted. Apparently the number of students will grow substantially in the coming years. There are 1200 faculty members in the University, and the well-known, Columbia-trained quantum chemist Professor T'ang Ao-ch'ing is the President of the University. Professor T'ang was the leader of the Chinese chemistry delegation to the United States in 1977.

Department of Chemistry. The Department of Chemistry has 520 students studying in six specialties: inorganic, organic, physical, analytical, polymer, and biochemistry. The number of students, including graduate students, will increase to 850 in the summer. Among the 220 teaching staff in the department, only 17 have the rank of full or associate professor; 28 are lecturers and the rest are teaching assistants. This skewed distribution of ranking is mainly due to the removal of the ranking system during the Cultural Revolution; however, it is expected that many teaching assistants and lecturers will be promoted in the near future.

There is quite active research work taking place in the department. The graph theory of molecular orbitals (HMO) is the most distinguished accomplishment. In addition, we heard presentations on the following subjects:

Molecular shell models and ligand field theory

Statistical theory of polymers on the problem of sol-gel distribution during chemical cross linking

The molecular weight distribution and kinetics of living polymers

Molecular sieve zeolites

The determination of trace total mercury in foods, soils, and sediment

A study of catalysts in heterogeneous oxidation of ethylene to acetaldehyde

Undergraduate analytical chemistry laboratories were very well equipped. This department has the responsibility of organizing an analytical chemistry course for Chinese universities. Although there were several setbacks in both teaching and research under the Gang of Four, this department seems to be thriving and gave us a very favorable impression.

Instrumentation available in laboratories for third-year students include a visible-UV spectrograph of Chinese design similar to a Beckman DU, 1-meter arc emission spectrometers, a variety of GCs all of Chinese manufacture, a gel permeation chromatograph of Chinese manufacture, a BET device, and a well-equipped semiconductor laboratory.

22. Kirin Institute of Applied Chemistry, Ch'angch'un

The Kirin Institute of Applied Chemistry was established in 1948 after Liberation as a multibranched institute. In 1953 part of the institute combined with the Institute of Physical Chemistry in Shanghai and the remainder was organized as the Institute for Applied Chemistry. The Director is Wu Hsueh-chou, a spectroscopist who is an alumnus of Cal Tech. The Associate Director is Ch'ien Pao-kung, a polymer physicist who had worked with Overberger at Brooklyn Polytechnic (1946–49).

Total personnel in the institute is about 1000, about 600 of whom are in the laboratory. The latter include 6 research fellows (equivalent of professors), 31 associate research fellows (equivalent of associate professors), and 68 assistant research fellows (equivalent of assistant professors). There are approximately 300 bachelors-equivalent personnel. These are not considered to be of adequate scientific training and quality, which is clearly a result of the low level of scientific training received by students in the universities during the Cultural Revolution.

Research is carried out in the following areas:

1. High polymer chemistry and physics
   - (a) synthesis of olefins and diolefins, and the use of transition metal and rare earths as catalysts in polymerization
   - (b) synthesis of heterocyclic polymers
   - (c) radiation chemistry of high polymers including cross-linking
   - (d) high polymer physics, which includes structures and mechanical, electrical, and solution properties of high polymers.

2. Structural chemistry (molecular spectra, X-ray diffraction, NMR, ESR, and mass spectrometry)
3. Catalysis (heterogeneous oxidation catalysis, homogeneous catalysis)
4. Electrochemistry, including metal corrosion
5. Laser chemistry and laser spectroscopy (using home-built laser equipment)
6. Inorganic chemistry and inorganic analysis
   (a) rare earth chemistry (extraction, separation, and synthesis of rare earth compounds)
   (b) photoelectric properties
   (c) analysis of complex ores

Instrumentation. Instrumentation at the Kirin Institute of Applied Chemistry includes a 60-MHz JEOL NMR, a JEOL mass spectrometer model DMS 100, infrared, UV, and gas chromatography. They have a home-built 7-watt argon ion laser powering a dye laser, and a TEA-mode CO_2 laser.

In the NMR area they are making applications of NMR to structural analysis, and they are doing proton spectra and measuring relaxation times. There is also an instrument development program. They are designing their own 100 MHz-FT spectrometer, which will use a DTS-130 computer. Part of the goal of this program is to become self-sufficient. They are having difficulties with the field homogeneity in the magnet they have built, which goes to 23 kG.

Some additional instrumentation available for research at this institute includes a torsion braid analyser of U.S. manufacture, a Chinese-built stress–strain device, a square wave pulsed polarograph, a HPLC, a He–Ne laser light-scattering device, a plasma emission (1 meter, Hilger-Watts) spectrograph, two conventional X-ray sets with Phillips' cameras, a torsion pendulum mechanical spectrometer, and a Hitachi SEM.

23. New China Printing Plant, Ch'angch'un

This factory started production in 1943 during the Japanese occupation. Improvements have been made over the years, but there were setbacks during the time of the Cultural Revolution and the Gang of Four. There are now 1600 workers, and they print more than 6000 tons of paper per year. They print the works of Marx, Lenin, and Chairman Mao; Red Flag material; propaganda; pictures; portraits of Chinese leaders; dozens of journals; and a large volume of textbooks for students. Printing of textbooks for students is their present main task.

There are four workshops in the factory for typesetting, printing, color printing, and binding. They also have clinics, nurseries, and other welfare departments.

Innovation is accomplished by their experienced workers; there are no engineers. About 80% of the work is done by machinery in an automatic manner. The workers have invented machines for such assembly line operations as turning the pages and doing the finishing steps in the binding process. We found it particularly interesting to watch the men and women select type from shelves containing thousands of characters to typeset in Chinese. Of special interest was a brand new typesetting machine (much like a Chinese typewriter) that uses a photoreduction method, operated by a very skilled woman.

Our visit took place on a Sunday and the plant was in full operation. The day off for Chinese workers varies by district and happened to be Friday for this factory. Our host presented each of us with a lead-tin printing piece with our name on it, and we were given some large posters with color pictures of Chairman Mao with Chairman Hua and color posters used to advertise the Chinese Spring Festival.

24. Refinery Complex No. 2, Fushun

The Fushun Petroleum Refinery originally produced fuel from shale oil. In 1952, the plant was converted to a crude oil refinery. It had an original capacity of 1,500,000 tons/year of crude, and this capacity has been raised to a current level of 4,000,000 tons/year. The crude is transported from Tach'ing to Fushun via heated pipelines.

The refinery has an atmospheric distillation column, a vacuum distillation column, a dense fluid bed catalytic cracker using zeolite catalyst technology, a fixed bed platinum reformer and a delayed coker. The fixed bed platinum reformer is of Italian design and construction (SNAM) and still employs Italian platinum catalyst technology. The other units are Chinese-designed and built. There is also a sulfuric acid plant and an electric power station. The refinery produces a range of fuels and chemical products, as shown in Table VIII. About 5% of the crude is converted to gasoline, 12% to diesel oil, and the remainder is used as industrial fuel. The specifications of this refinery are also given in Table VIII.

Shale Oil Facility. The first Fushun shale oil plant was constructed in 1941 during the Japanese occupation. During the 1931–45 period, the Japanese mined 74 million tons of shale rock. The shale is obtained from the nearby coal mine and is mined concurrently with the coal. There are two shale plants currently operating in Fushun, one with a production of 80,000 tons/year of oil and the other with a production

TABLE VIII
FUSHUN REFINERY

Crude Source: Tach'ing
Crude Run: 4,000,000 tons/year
Supply Route: Heated pipeline
Plant Area: 2.5 km^2
Workers/Staff: 8700

Process	Capacity, tons/year	Technology
Petroleum Refining		
Atmospheric Distillation, Vacuum Distillation	2,800,000	Chinese
Fluid Bed Cat. Cracking	900,000	
Fixed Bed Reforming	100,000	Chinese—Zeolites
Delayed Coking	400,000	Italian (SNAM, Pt/Al_2O_3)
Chemicals		
BTX from Fixed Bed Reforming		
Ethylene from Catalytic Cracking		
Fushun Products		
Kerosene		
Jet Fuel		
Gasoline (76 to 95 O.N.)	5% of crude	
Naphtha		
Diesel Oil	12% of crude	
Fuel Oil		
Soap Wax		
Detergents		
BTX		
Ethylene Glycol		
Lube Oil		
Diethanolamine		
Monoethanolamine		
Propylene Oxide		
Ethylene Oxide		
Isopropanol		

of 120,000 tons/year of oil. There is another plant that produces 80,000 tons/year of shale oil close to Canton. The process itself consists of a series of rotating retorts fired by the partial burning of oil shale rock, which drives off the volatiles as a mist. The shale is crushed and fed into the top of the furnace, the rock travels down through the furnace and after a nine-hour residence time the spent shale leaves from the furnace bottom. The volatiles are passed through a water wash to remove the oil, which is separated from the water by open ponding.

Ammonia is scrubbed in dilute sulfuric acid and obtained as ammonium sulfate. About 5 wt % of the shale is converted to oil.

The shale oil is apparently not processed at the Fushun Refinery. Instead it is shipped to local users. These nonspecified local users were said either to burn it directly or to "convert" it on site for individual needs.

25. West Open Pit Coal Mine, Fushun

This huge open pit mine excavates coal and oil shale. Currently the production rate is 3.6 million tons per year of coal and 12 million tons of shale rock per year. The pit is 6.6 km long and 2.2 km wide, with a current depth of mining of 270 m. The mine was started in 1914. There is a deep shaft mine in another part of Fushun as well as another open pit mine. There are in each of the mines tertiary and quarternary rocks that are 1 to 50 million years old. We were told by the Chief Engineer that there is a depth of 30 m of soil, 300 m of a green rock layer, 110 m of an oil shale layer, and finally a coal layer of 120 m before igneous rock is reached. With these figures, the final depth of the mine should be 560 m. These numbers do not seem to be consistent with a total depth of 270 m. Later on it was pointed out that the thickness figures were measured not as a vertical thickness but as an incline thickness, with the angle of inclination between 17° and 19°.

26. Institute of Organic Chemistry, Shanghai

The Shanghai Institute of Organic Chemistry, founded in 1950, is probably the most active center for organic chemistry research in China. It moved to new quarters in Shanghai in 1959. This institute has as its Deputy Director Huang Wei-yuan, an organic fluorocarbon chemist who was educated at Harvard (Ph.D., Fieser, 1952) and who was a member of the Chinese Chemistry Delegation to the United States in 1977. Ms. Hsia Tsung-hsiang, an organic analyst at this institute, was also a member of that delegation.

This remarkable institution occupies several buildings, and an additional laboratory is planned for construction in the near future. There are currently approximately 500 research workers at the institute, two-thirds with a bachelor's degree or equivalent training. Twenty-six of the staff are considered to be of professorial level. In addition, there are approximately 200 other supporting workers in clerical and technical services. Research work is divided into three general areas: natural products, organoelement chemistry, and physical organic chemistry.

High pressure liquid chromatographs are fabricated in the institute shops. Research facilities include infrared, 60 MHz NMR, mass spectroscopy, electron microscopy, and X-ray diffraction. An imported ESCA spectrometer is also reported to be available in this institute. Some of the other instrumentation that is available for the support of the research at this institute include a vacuum coating apparatus of Chinese manufacture, various homemade HPLCs of good quality with a flame ionization detector using a moving wire and a UV detector employing a FET on a photosensitive element, a novel H/C analyzer (see Section 6-F), a Cl^-, Br^- automatic titrator, a CO_2 laser for pyrolysis, and a differential scanning colorimeter.

27. Institute of Biochemistry, Shanghai

The institute of Biochemistry in Shanghai has just completed a new seven- or eight-story building to house the 200 research and technical staff that includes 150 people with B.S. degrees. The present structure of this institute evolved in 1958. The work is divided into a number of areas: (1) protein chemistry and polypeptides with emphasis on synthesis, (2) synthesis of polypeptide hormones, (3) structure and properties of insulin, (4) oligomeric nucleotides, (5) enzyme chemistry including structure and properties, kinetic studies, biomembranes, and solid-state enzymes, (6) hepatoma and its early detection and relation to hepatitis, (7) biologically active enzymes, (8) plant viruses, and (9) genetic engineering, which is just beginning. Other areas in which the institute is involved include a small factory of about 100 workers for the production of more than 500 chemicals including enzymes, coenzymes, amino acids, and nucleotides and an instrument laboratory in which ultracentrifuges, automatic amino-acid analyzers, and similar instruments are built.

The demonstrations or discussions we had dealt with insulin synthesis, structure and mechanism of action; glucogen synthesis; and a theoretical discussion of the Eigen-Alberty-Hammes enzyme reaction model.

The only instrument of special interest was a Jasco G-20 CD-ORD machine. Other instrumentation available to support the research work at this institute includes automatic amino-acid analyzers of Chinese manufacture.

28. Institute of Nuclear Research, Shanghai

This institute was established in 1959. Its Director, Chin Ho-tsu, has been with the institute since its founding, but he was not present for Seaborg's visit in 1973 nor for the present visit. The Deputy Director is Shih Hsuan-wei, who also was not present during our visit. The person in charge of our visit was Chao Chung-hsien, Chief Engineer and principal host. Professor Seaborg also met him on his previous visit. Another in the host party was Dr. Liu Nien-yen, who is in charge of the radiochemistry and isotope department.

The institute is organized into five sections:

1. Nuclear physics
2. Isotope research
3. Radiation chemistry
4. Nuclear electronics and detectors
5. Accelerators

They have a ^{60}Co source of 120,000 curies for radiation research, purchased from AECL (Canada), as well as a 1.2-meter diameter cyclotron that was built in the institute. They intend to convert this to a sector-focused machine with a capability for 10–30 MeV protons, 20–60 MeV deuterons, and heavy ions. There are plans to build a Tandem accelerator of total energy of 12 MeV for protons. This accelerator should be similar to the FN accelerator that is marketed by High Voltage Engineering Corporation. The accelerator is currently in the design stage, and a scientific program for the use of the accelerator is in the process of being set up. It is anticipated that the work with the Tanden accelerator will be application-oriented, and it is not intended to be a forefront instrument.

Some of the instrumentation available to support the research work at this institute includes a homemade perturbed angular correlation device, a proton X-ray fluorescence system, a Cockroft–Walton accelerator for the production of 14 MeV neutrons and tritium-labeling facilities.

Library. There is a new large library that was built three years ago. In addition to journals, it has a holding of 100,000 books. The size of the library is justified on the basis that the institute is remotely located from Shanghai.

29. Institute of Materia Medica, Shanghai

The Institute of Materia Medica, Shanghai, was founded in 1932 and includes departments of synthetic organic chemistry, phytochemistry, analytical chemistry, antibiotics, and pharmacology. There is also a pilot plant at this institute. Among the projects being pursued is extensive work on camptothecin and various derivatives.

Some of the instrumentation available to support the research work at this institute include a 100-MHz JEOL NMR for H^1, F^{19}; a homemade HPLC that is different from the Institute of Organic Chemistry version; and toxicology assessment facilities that include laminar flow hoods and electrophysiological facilities.

30. Futan University, Shanghai

Futan University was established in 1905 and has 14 departments, including both the arts and the sciences. The sciences include mathematics, physics, chemistry, zoology, optics, computer science, and nuclear physics. The university currently has 3600 students and 2100 faculty. The chemistry department has 240 students and 200 faculty, including 40 in the professorial ranks. Lu Ho-fu, Professor of Theoretical Physics at Futan University, did his graduate and post-doctoral work with John Tate during the period 1936–41 at the University of Minnesota, where he determined the abundance ratio of the boron isotopes. In addition to the usual specialties in chemistry, there is a workshop and a small plant for polymer production. The areas of research covered by the department are organic, inorganic, analytical, and physical chemistry; catalysis; electrochemistry; and polymer science.

Teaching Laboratories. The goal of this laboratory program is to educate students in instrumental analysis, to train them in the design of instrumentation, and to give them exposure to research in environmental chemistry. The course is offered to current third-year students. The program is very well organized and has some excellent equipment. The academic content of the program was significantly interrupted during the 10-year reign of the Gang of Four, but the equipment remained intact during that entire era. Students are trained in optical methods of analysis, electrochemical methods of analysis, and chromatography. In the area of optical analysis, they have ultraviolet and visible spectrophotometry and emission spectroscopy. They hope to have infrared absorption spectroscopy by 1980. In the area of electrochemical analysis, they use coulometry, electropotential analysis, ion-selective electrodes, potentiometric titrations, polarography, and anodic stripping voltametry. Work in chromatography is limited to gas chromatography using both thermal conductivity and flame ionization detectors. They also do some work in atomic absorption and have a Zeiss (Jena DDR) atomic absorption spectrometer and another one that was made in Shanghai.

Students are encouraged to get involved in small research projects. Some of the recent projects carried out by students include determination of metal ions in fish and rice by ASV and the determination of mercury vapor in the air by cold vapor atomic absorption.

The person in charge of this program is Professor Ku Yi-tung, who received a Ph.D. with Stieglitz at the University of Chicago in 1935 (he knew the late Herbert N. McCoy).

The teaching laboratories in polymer science for third-year students seem to be excellent. The laboratory contains a Chinese-designed tubular pyrolysis gas chromatograph for analysis of copolymers (said to give analysis in half an hour with 2% accuracy on a 2-mg sample). A unique experiment was the use of "inverse gas chromatography" for studies of T_g's of blends by Ho Man-chün and Chang Chung-ch'uan. The probe vapors were benzene, toluene, heptane, ethyl alcohol, or cyclohexane. Polycarbonate/polystyrene blends were studied to determine the range of compatibility. Blends containing less than 7% PS are found to be compatible at 175°–200°C (MW of PS = 150,000; MW of PC = 180,000).

Futan University possesses a DSC and a GPC of the type designed at the Peking Institute of Chemistry. A second very simple GPC for student course use has been constructed. There is also a homemade torsion pendulum for mechanical studies involving an optical system employing a photodiode permitting the recording of the vibrations with a strip-chart recorder. It was commented that data recording was time-consuming, and there was hope of using computer analysis in the future. High-impact polystyrene samples containing two sizes of rubber particles were studied and two loss peaks corresponding to these were found (suggesting two separate T_g's). A homemade gas permeability apparatus was in use for studying N_2 permeability of polymer films of interest for food packaging (Saran, PE, PP). A density gradient column was employed for PET sample density measurements utilizing a xylene/CCl_4 gradient. Floats are not commercially available and must be locally made and calibrated. There was also a homemade thermomechanical analyzer involving a homemade linear variable differential transformer (these may now be obtained based upon a design by the Peking Institute of Chemistry).

The teaching laboratory in polymer science for undergraduate students is excellent and is comparable or superior to the laboratory courses available in the United States to seniors or first-year graduate students at the very few universities where such courses are offered. The laboratory suffers from the lack of modern instrumentation, but many ingenious instruments have been improvised, most of which are quite adequate. In some respects the pedagogical value of using such homemade instrumentation may be greater than the use of the commercial "black boxes" prevalent in U.S. laboratories, where often the student does not appreciate what goes on inside. Courses in polymer chemistry, polymer physics, and polymer technology are available for third-year students. There does not seem to be much interaction between the catalytic polymerization efforts of the catalysis group and the polymer group.

Library. The library of Futan University is quite good. The foreign periodical collection in polymer science is excellent. Journals (or reproductions of the original journals) such as *J. Polym. Sci.* are about six months behind publication date. Original copies of chemical abstracts, three months behind publication date, are available. Back issues are fairly complete, but foreign book holdings are deficient.

Instrumentation. Instrumentation available to support the teaching and research programs at Futan University include a Rigaku 12-kW rotating anode X-ray source for powder and low-angle work, a GC of Chinese manufacture, a Japanese EM, a GC with tubular oven for copolymer sequence analysis, a gel permeation chromatograph, a diffraction scanning colorimeter, a Mettler P-160 N balance, and a Van de Graaff accelerator.

31. Shanghai Institute of Chemical Engineering

The Shanghai Institute of Chemical Engineering was established in 1952 as a college of chemical technology. It combined five existing chemical engineering departments from Eastern China. Of the five universities, four were from Shanghai: Chiaot'ung, Tat'ung, Ts'untan, Fuch'iang; the fifth, Tungwu, was from the city of Soochow. Since it is a freestanding chemical engineering college of the Soviet model, it has to be self-sufficient. It is divided into petroleum chemical engineering, organic chemical engineering, inorganic chemical engineering, chemical machinery, and fundamental training (which includes mathematics, chemistry, physics, and languages). It contains several research laboratories and has a staff of 2400, including a teaching staff of 990 of whom 68 are professors and associate professors. Professor Wang Chen-ming studied with Hauser at MIT and worked at Lummus in the United States for several years. Professor Li Pang-chen studied with Peters at the University of Illinois, Urbana. The campus has an area of 500,000 square meters with 130,000 square meters of building space. It has 400,000 books and 1200 journals in the library. Since its founding 13,000 students have graduated.

The Shanghai Institute of Chemical Engineering currently has 2600 students in a three-year program that is being changed to a four-year program. Examinations have been reintroduced with emphasis on the theoretical and fundamental, as well as on laboratory courses. In the 1978 examination all the spaces were openly competed for by high school graduates and past students. There are only five graduate students currently enrolled, but 55 new graduate students will be admitted in the fall. Graduate students will go through a three-year program, at the end of which they will not receive degrees but will receive certificates. At one time they had a scientific committee to help decide which research projects to undertake; this committee is not operating now but is supposed to be started again.

The chemical engineering curriculum is currently a three-year program. In the first year the curriculum consists of mathematics, physics, inorganic chemistry, analytical chemistry, drawing, and foreign languages. The second-year curriculum consists of mathematics, languages, mechanics, organic chemistry, physical chemistry, and electrical engineering. The third-year program consists of chemical engineering, automatic control, polymer chemistry, and technology. We did not get a breakdown on what the chemical engineering courses are.

32. General Petrochemical Works, Shanghai

The Shanghai Petrochemical Complex is a chemical refinery that uses a gas oil steam cracker to produce olefins and aromatics. These are converted on site into plastics and fibers. Tach'ing crude is used as feed for the refinery and is delivered in heated tankers by sea to Ch'inghuangtao, an oil receiving port some 25 km west of the city. The oil is heated before loading at Tach'ing and the tankers are insulated. At the port the oil is reheated and piped to the refinery. At the refinery the crude undergoes simple atmospheric distillation to separate the light materials and the gas oil is distilled out and steam cracked. The bottoms are used as fuel to generate electricity and as fuel in other parts of the refinery. Naphtha is sent to a fertilizer plant a few kilometers away and used as feed for ammonia production.

After gas oil cracking the ethylene is separated and used to produce high-pressure, low-density polyethylene. It is also oxidized via the Wacker $PdCl_2/CuCl_2$ technique to yield acetaldehyde, which is oxidized to acetic acid and used to produce polyvinyl alcohol and ultimately vinylon. The plant produces vinylon fiber. Propylene is ammoxidized to acrylonitrile, which is then used in the production of polyacrylonitrile. The aromatics are sieved to produce a *p*-xylene fraction, which is oxidized to TPA and converted to dimethylterphalate (DMT) via esterification with methanol. The alcohol is shipped into the refinery. The plant produces terylene fiber from the DMT. The *o,p*-xylenes are isomerized and recycled to extinction. Some *trans*-alkylation is performed yielding monoethyl and trimethylbenzenes using a zeolite catalyst.

The Chinese expressed a desire to begin nylon-66 production and had a team in the United States discussing hydrocracking and reforming technology. The aim here is to hydrocrack to produce a saturated fraction which can be performed to yield an aromatic cut useful in nylon-66 manufacture.

Plant History. This plant was conceived and designed at the Shanghai Chemical Industry Design Institute, which reports to the Ministry of Chemical Industries. Construction of the plant was begun in 1972 with the building of a dyke to reclaim some land from the sea. The first pilings for the plant were sunk in 1974 and the first units came on stream in 1976. Most of the technology employed is foreign as noted in Table IX. The same strategy was employed here as at the Peking Petrochemical Works. The Chinese jointly managed on-site construction of individual units with as much Chinese-constructed machinery as practicable. They are planning further construction, including a nylon-66 plant and a research institute.

The total site includes a workers' village with apartment and store complexes, hospitals, primary and middle schools, and water treatment facilities. This complex houses some 60,000 people of whom 24,000 work at the refinery, including 10,000 workers, 1000 technicians, and 100 engineers. The refinery could use 2000 additional engineers but only manages to obtain one out of every three engineers it requests. The plant management considers the ratio of engineers unfavorable and plans to initiate local schooling with emphasis on petroleum chemistry at the high school and college level.

Refinery management is split between Peking and Shanghai. Production responsibility is centered at Peking in the Textile Ministry, while local administrative responsibility is centered in Shanghai at the Industrial and Transportation Committee.

Salaries at the plant vary between 80 and 300 yuan (48 to 180 U.S. dollars/ month), and everyone gets company housing. The housing costs an average of 4% of salary and space is based on need, not rank. For example, a couple with children would have more space than a childless couple.

In terms of recruiting, the plant manager requests personnel from the Education Bureau in Shanghai. The Education Bureau receives information on candidates from different universities in the form of complete dossiers. The bureau then decides on priorities and

TABLE IX
Shanghai Petrochemical Works

Crude:	Tach'ing, shipped via heated tanker and offloaded at an oil terminal located 56 km from the Petrochemical Works.
Capacity:	1,700,000 tons/ year
Location:	75 km southwest of Shanghai
Size:	10,600 acres 10 units, 2 plants (Chem. Plant 1 and Chem. Plant 2)
Workers:	24,000
Technicians:	1000
Engineers:	100
Electrical Generation:	250,000 kW 150,000 kW used in the refinery 100,000 kW used externally

Process	Capacity, tons/ year	Technology
Atmospheric Distillation		Chinese
Gas Oil Steam Cracking	470,000	Japanese (Mitsubishi)
Polyethylene	60,000	Japanese
Acrylonitrile		Japanese built, SOHIO catalyst
Polyacrylonitrile	47,000	
Acetaldehyde	30,000	
Acetic Acid	30,000	Acetic acid oxidation (acetate), Chinese
Polyvinyl Alcohol	33,000	Japanese
Terylene	25,000	West German

the college graduates are assigned without plant trip interviews. The promotional system for new college entrants has three steps. At the beginning the graduate is called an apprentice and within one year is generally promoted to technician. After that promotions are based on the recommendation of his group head and co-workers. Criteria used include ability, training, diligence, and political attitude. The third level is engineer.

33. Institute of Ceramic Chemistry and Technology, Shanghai

The Institute of Ceramic Chemistry and Technology was formerly a part of an Institute of Metallurgy and Ceramics of the Chinese Academy of Sciences. In 1960 the Institute of Metallurgy and Ceramics was separated into two institutes, the Institute of Metallurgy and the Institute of Ceramic Chemistry and Technology of the Chinese Academy of Sciences. The latter currently employs about 900 people, including 500 scientific and technical personnel, 300 staff in shops and pilot plants, and 100 administrative personnel.

The Director of the Institute of Ceramic Chemistry and Technology is Yen Tung-sheng, who received his Ph.D. in chemistry from the University of Illinois with Professor Hersch during the period from 1946–50. He indicated that the name of the Institute does not accurately reflect the nature of the work that is being carried out. The institute includes a workshop and a department for designing equipment—both electrical and mechanical. Most of the research at the institute is for the long-range needs of electronic and laser applications.

The Institute is beginning to accept graduate students. Of the approximately 100 students who indicated first priority interest in this institute at the preliminary examinations at the end of June, 16 or 18 students were to be accepted for a final examination to be given at the Institute. The Institute planned to take 9 or 10 students beginning in the fall of 1978. These would be the first students in about 10 years.

Instrumentation available to support the research program at this Institute include X-ray fluorescence and X-ray diffraction, emission spectroscopy, atomic absorption spectroscopy, an electron microscope of East German origin, a Chinese-made He–Ne laser system for studying acoustic–optical materials, a laser doubling system with a YAG pumping laser, a vacuum evaporation apparatus for nonreflection coatings, a Rigaku 30-kw (600 kV, 500 mA) rotating anode used as an X-ray source for low-angle scattering, an Ar cw laser (4880 Å), a He–Ne laser for holography, a Zeiss (Jena) phase contrast microscope, and several other optical microscopes.

34. Shanghai Normal University

Shanghai Normal University was founded in 1951 as an amalgamation of parts of or all of some universities that ceased to function after Liberation in 1949. With a present faculty of 1100 teachers, it is devoted to training students for teaching posts mainly in middle schools, but some are trained for lower level schools and others are qualified to teach basic courses in colleges and universities. The curriculum has been confined to three years with about 4000 students, but it is now planned to expand to four years with up to 5000 students. The student body at present includes about 100 graduate students enrolled in a three-year curriculum that, upon completion, qualifies them to teach at a higher level or to accept positions at research institutes. The university also prepares teaching programs for Shanghai TV stations, which are viewed by some 2000 students.

There are 10 departments: Chinese, foreign languages, history, pedagogy, political education, mathematics, physics, chemistry, biology, and geography. In addition there are three research units (or institutes): estuary and coastal, foreign education, and world history. They plan to increase the number of research units to strengthen their teaching program in science and technology.

Shanghai Normal University, in connection with its function of training teachers, also has schools at the kindergarten, elementary, and middle school level associated with it. All students go to the middle school, reputed to be the best in Shanghai, for six weeks of teaching practice.

The recently instituted national entrance examination is now being utilized to choose students. The students are graded on a system of 1 to 5, with 3 and above passing. If a student fails (grade below 3), the course is repeated; the student cannot, in general, fail, although this could happen on the basis of attitude or behavior. There are seven 50-minute teaching periods per day, four in the morning and three in the afternoon, six days per week.

The chemistry department occupies about 6000 square meters of floor space for classrooms and laboratories. In addition to courses in foreign languages, politics, pedagogy, physical education and teaching practice taken by all students, the prospective chemistry teachers take courses in inorganic chemistry (first year), organic chemistry (second year), analytical chemistry (second year), and physical chemistry (third year). In the now added fourth year, they will take a course in structural chemistry in the first half and will undertake a special research project with a written report in the second half. In addition, chemistry students take courses in physics and mathematics, and in the fourth year, a course in chemical engineering. They are translating the textbook, laboratory manual, and teacher manual of Chem Study into Chinese for use in their courses. The Chem Study films are, as yet, not available to them.*

In addition to the main library each teaching department has a library. Their heritage from St. Johns University, disbanded following Liberation in 1949, includes a set of *Chemical Abstracts* starting with Vol. 1 in 1907. The Chemistry Department Library is provided with about 140 chemistry journals and has some 5000 foreign language books, while the main library is provided with some 1000 journals.

* A representative set of these films was sent to them after the delegation returned to the United States.

35. Shanghai Water Works

A visit to the Shanghai Water Works focused on discussion of water quality, standards, and new trends in treatment. The turbidity of treated water for use in Shanghai was 2.4 units, and color was (15), both indicators of poor filtration. We recommended consideration of use of polyelectrolytes, dynamic feed of alum, and increased pH for control of corrosion and solubility of Cd, Cn, and Pb.

36. Computer Technology Research Institute, Shanghai

This institute was founded under the auspices of the Shanghai Academy of Sciences in 1969 with less than 50 staff members. It was originally called the Computing Center of Shanghai and its mission was to service a computer called the X-2. The same group of people collaborated with Futan University in designing and constructing the prototype of the 709 computer. Work on this computer was started in September of 1970 and it was completed by December 1971. In 1973 the current institute was formed. Its mission is to maintain and design computers, to do research on computational methods, and to carry out some scientific computing. Currently the institute uses the 709 computer, and a more advanced 731 system, while the old X-2 computer has been given away.

37. Shanghai Industrial Exhibit

The Shanghai Industrial Exhibit opened in 1969. It consists of a number of exhibits of industrial items manufactured mainly in the Shanghai area, all of which are supposed to be available for sale. The exhibition hall covers some 50,000 square meters in several buildings with 10,000 square meters actually used for exhibition purposes. The exhibits range from coal-fired 1000 #/hr steam boilers to large earth moving tractors. The following chemical products were noted:

> Ammonia; ammonium bicarbonate, said to be manufactured by both local small plants and a few large central ones; and NPK fertilizers, stated to be in experimental production stages in Shanghai but being produced elsewhere in China.
>
> High-activity and low-toxicity pesticides and new antibiotics for rice blight being produced at the Shanghai Agricultural and Chemical Factory.
>
> Polymers, including polyethylene, reinforced nylon-10-10,

polycarbonate esters, polypropylene, Dacron, Orlon, polyamides, lucite, nylon-6, Teflon, ABS, polycarbonates, and polysulfanes.

Drugs, including insulin, Vitamin B, birth control pills, and herbal extracts that in combination with Western drugs control hypertension.

Fluorescent powders used in TV tubes.

Metals, including platinum, gold, silver, ytterbium, and palladium produced as by-products of the smelting and electrorefining of copper; titanium alloys used in rocket production; and zirconium cladding used in nuclear fuel shielding.

Rare earth metals that are found in northeast China including: cerium oxide and nitrate, samarium, scandium, ytterbium, and europium. Shanghai is the center of production, which began in 1960.

Silicon solar cells also made in Shanghai.

Exhibit personnel were not conversant with technical details, so there is some uncertainty about the accuracy of the information. There was an exhibit of large tractor tires and industrial rubber belting. These were said to be produced from a blend of natural and synthetic rubber and use nylon, rayon, or steel cord. Silicone rubber tubing for medical applications was also shown. An exhibit of petrochemical products featured polymeric materials made at the Kaoch'iao Chemical Plant near Shanghai and included polyethylene, polypropenes, polystyrene, polycarbonate, polymethylmethacrylate, ABS resin, and "synthetic wool" of polyacrylonitrile. Other plants in the areas were said to produce reinforced nylon-10-10, polysufones, polymide, teflon films, and fibers of Orlon (PAN), Dacron (PET), vinylon (PVOH), and nylon-6. While close examination of products was not possible, they superficially appeared to be of reasonable quality.

An impressive injection molding machine was exhibited as well as a large array of textile spinning, weaving, and knitting machinery, which, on the basis of casual observation, was of comparable quality to such machines in the West and similar to those seen in actual use in the Peking vinylon plant visited.

The computer displayed, a DJS-131-1, was not impressive by modern standards. It processes at 500,000 cycles/sec and has 32 K internal storage. It uses punched tape input (no magnetic tape) and has a printer and XY plotter output. It uses ALGOL (no FORTRAN). It is fairly large in size by U.S. standards. It is not much more advanced than the computer in use at the Shanghai Institute of Chemical Engineering.

38. Shanghai Refinery

This refinery was not visited, but we were told of its existence at the Shanghai Petrochemical Works. It has a crude run of 5,000,000 tons/year and uses Tach'ing crude. The refinery has at least the following units: fluid bed catalytic cracking, reforming (150,000 tons/year) and a vacuum still.

39. Chekiang University, Department of Chemical Engineering, Hangchow

Chekiang University is a school of science and engineering only, while subjects in the arts are taught at a different university in Hangchow. At one time the university had up to 11,000 students, but now they have only 3000 students and a teaching staff of 1600, 32 of whom are professors and associate professors, 400 of whom are lecturers, and the rest of whom are assistants. Two members of the faculty studied at MIT: Chen Chia-yung with Hottel and Williams and Chou Chun-hun with Vivian. Traditionally one-fifth to one-fourth of all the students have been in chemical engineering, which is by far the strongest department in the university. They are currently admitting 2500 new freshmen, which will presumably build up the total student body.

The Department of Chemical Engineering was founded in 1928 and is the oldest in China. It is divided into physical chemistry, organic chemistry, analytical chemistry, and chemical engineering. A brief look at the laboratory showed a project of ethylene oxidation over a silver catalyst, an equilibrium study of gas and liquids, and experiments using other small-scale apparatus. The laboratory also has experiments in fluid mechanics, gas–liquid reaction, bubble column reactors, and an apparatus for automatic control of the level and temperature in a mixing tank. Most of the equipment is not adequate for research and is used for teaching undergraduate courses. All students either do a thesis or work at a factory as an equivalent experience.

They also have a section on chemical machinery, which is a subject not taught in the U.S. chemical engineering departments. Some of the problems that they study are the rupture of tanks and breaking of centrifugal parts. They have a laboratory of catalysis with BET and chromatography, but no computers.

There are no electives at this university and there are quite a few required courses. In chemical engineering, the following courses are required: 1 term of thermodynamics; 2 of transport phenomena;

1 in reaction engineering; 1 each in polymer chemistry, technology, and physics; 1 course in petrochemicals, which is mainly applied thermodynamics; 1 course in technology and equipment; 1 in design principles; and finally, 2 courses in unit operations laboratory. Chemistry requirements are inorganic chemistry, organic chemistry (2 terms), analytical chemistry (2 terms), and physical chemistry (2 terms). Physics requirements are 2 terms, and mathematics requirements are 2 terms. In addition they study the politics of Marx and Mao and have 6 semesters of English, physical education, mechanical and electrical engineering, and drawing. The three options in this department are chemical and petrochemical engineering, automatic control, and chemical machinery.

40. Northwest University, Sian

This university was established in 1937 and before 1949 had 500 students. Since 1949 it has expanded and now there are 10 departments. Seven of these are in the sciences or mathematics. They include mathematics, chemistry, physics, geography, biology, geology, and chemical engineering. There are three departments in the arts, including Chinese history, Chinese literature, and politics. There are 780 teachers, which include the ranks of professor, associate professor, lecturer, and assistant. At the time of our visit, there were 1700 students consisting of both graduate students and the four-year university students. The graduate students, who are called research students, do not do any teaching. There are also 750 staff, including the university president and administrative officers, responsible people of the departments, librarians, machinists, and other workers. Promotion in the ranks is supposedly based on the triple criteria of teaching, research, and service. Since research was greatly curtailed during the period of the Gang of Four, it may play a much smaller role for the long-overdue promotions at this time. There is a scientific committee that decides on promotions. The committee consists of professors and other workers; the promotions suggested have to be approved by the university president. To obtain infromation on ability in teaching, opinion is solicited from the students. For the service contribution, political activities are very important. Furthermore, an appointment to lecturer only needs approval from the university president, but an associate professor needs approval from the provincial educational bureau, and a professorial appointment must also be approved by the education ministry in Peking. Assistants in China are long-term and full-time employees.

The Chemical Engineering Department is part of Northwest University, a comprehensive university consisting of arts, science, and engineering. However, it is the only engineering department in the whole college of engineering. The other departments of engineering are at the Northwest Industrial University. The Chemical Engineering Department started in 1937 as part of Northwest Engineering College, which subsequently became divided and the Chemical Engineering Department became part of Northwest University in 1972. Currently there are 94 teachers in chemical engineering, two of whom are professors and the rest of whom are lecturers and assistants. There are also 200 students. In the future they hope to graduate 120 students per year.

The department is divided into five different subgroupings, the first two of which are the biggest. They are inorganic chemical technology, such as acid–alkali and ammonia synthesis, headed by a professor; organic chemical technology, such as polyvinyl chloride synthesis and petroleum cracking; unit operations, which include distillation, heat transfer, fluid mechanics and fluidization, headed by a professor; fundamental chemistry, which includes analytical, inorganic, and organic chemistry, mostly taken by first- and second-year students; and chemical machinery, including mechanics, strength of materials, analytical mechanics, machine design, electricity, instrument measurement, and drawing.

Despite the fact that they are in a comprehensive university with a department of chemistry, the students of chemical engineering do not now take chemistry courses there. However, they do take courses in physics and mathematics from the physics and mathematics departments.

The required courses are mathematics for two semesters to differential equations, two courses in physics, one course in general chemistry, two courses in organic chemistry, and three courses in physical chemistry. In chemical engineering, unit operations, fluid mechanics, heat transfer, reaction engineering, thermodynamics, chemical technology, and finally design or a thesis are required. Other requirements include foreign languages, the most popular being English or Japanese. Other topics include political economy (Marx and Engels and Chairman Mao).

On the problem of job-seeking for graduates, we were told that there is a great shortage of manpower in this part of the country and everyone has jobs. The Provincial Education Bureau jointly with the Examination Committee decides how many students to admit and into what areas they should go. Entrance is by a national examination. Job distribution is the responsibility of a university committee that consists of some professors and some workers. This committee looks over the students' dossiers and suggests an appropriate placement. These suggestions are then reviewed by the Provincial Education Bureau and the Provincial Chemistry Industry Bureau, which makes the final decision as to who should get which job. The chemical factories in the Sian area produce ethylene, chlorine, and caustic soda by an electrochemical method, polyvinyl chloride, and a substance they call 666, which we were not able to identify.

41. Institute of Chemical Physics, Lanchow

The Lanchow Institute of Chemical Physics began in 1958 as a branch of the Institute for Petroleum Research and changed its name in 1961. It employs approximately 600 people, 300 of whom have college degrees and of the remainder, 100 are in the workshop and 200 are in administration. The current work of this institute is divided into three areas: molecular catalysis, functional materials, and an area encompassing synthetic chemistry, analytical chemistry, and instrumentation. Projects in these various categories include:

1. *Molecular catalysis.* The main emphasis is on homogeneous catalysis, on supported catalysts and on immobilized enzymes. Specific topics that were mentioned were hydroformylation, particulary by triscobalt-ammonia carbonyl, the photoassisted catalytic redox processes, the role of reversible oxygen carriers in catalytic oxidation, and the immobilization of glucose isomerase. It should be pointed out that these are planned topics and that in many areas work has not begun.
2. *Catalysis by heterogeneous materials.* There is continued interest in a deoxydehydrogenation of butene to butadiene. New projects include metallic and metallic oxide active sites and their syntheses and correlation with reactivity, surface physics, kinetics, catalytic oxidation of ethylene, the structure and behavior of rare earth catalysts, and also single-step synthesis of isoprene from formaldehyde and isobutylene.
3. *Synthetic chemistry.* Work in synthetic chemistry will include the synthesis and use of crown ether complexes and cryptate complexes; the production of functional polymers; and the study of polymer membranes for selective gas permeability.
4. *Analytical chemistry.* Analytical work seems to be concerned with high-resolution separation methods and high-resolution detection methods with concentration on chromatographic, spectroscopic, and inorganic analyses.

In addition to the Chinese-designed prototype GC-MS shown in Fig. 21, there is a SEM prototype of Chinese manufacture. The GC-MS includes a low-resolution mass spectrometer with a capillary GC, jet interface, and a 4 K (16-bit words) computer with a 48 K magnetic drum, all of Chinese manufacture.

42. Institute of Modern Physics, Lanchow

Yang Ch'eng-chung, the Director of the Institute of Modern Physics in Lanchow, has served in that capacity since its inception. He spent from 1945–51 in Liverpool, England, where he worked with James Chadwick and received his Ph.D. degree. Upon returning from England in 1951, he joined the staff of a newly formed (1950) nuclear physics institute in Peking. This institute served as the source of three institutes devoted to nuclear science created in the latter half of the 1950s: the Institute of Atomic Energy located near Peking; the Institute of Modern Physics in Lanchow; and the Institute of Nuclear Research near Shanghai. Yang moved to Lanchow in 1957 and supervised the building of the institute and its facilities prior to its formal operational start an 1963.

The Institute, with a staff of more than 300 scientists and technicians, now has five research departments: nuclear physics; design of new accelerators; electronics, detectors, and computers; heavy ion (1.5-meter) cyclotron; and applied nuclear physics (including radiochemistry).

Research in nuclear physics features research on fusion and fission, transfer reactions, high angular momentum states, nuclear spectroscopy, and theoretical physics. The products produced in heavy ion reactions are identified by the use of radiochemical techniques.

In addition to a heavy-ion cyclotron, the Institute also has a Cockroft-Walton accelerator built in Shanghai in 1964 that has an operating voltage of 400 keV. This is used for the production of 14 MeV neutrons by the D-T reaction. Use of a rotating target makes possible the production of a yield of 3×10^{11} neutrons per second. This facility also makes available a beam of deuterons with which to conduct low-energy nuclear physics experiments.

The institute has been authorized to build a 6.15-meter heavy-ion cyclotron that will give them the capability to accelerate ions as heavy as xenon to energies of 6 MeV per nucleon. A previous program on the development of high intensity plasma sources at this site has been discontinued.

Figure 45. Peripherals at the Computer Center in the Institute of Modern Physics, Lanchow.

Some of the instrumentation available to support research work at this institute include coincidence spectrometers, a 16,384 multi-channel analyzer and multi-8 computer (32 K memory, 16-bit words) (Intertechnique, France), Tektronics scopes 581-A (1966) with type 86 plug-ins (obtained in a circuitous way), a Mössbauer spectrometer, and a computer center with a 64 K memory, 24-bit words, 2-μsec computer using FORTRAN II. Some of the peripherals to this machine are shown in Fig. 45.

43. Institute of the Lanchow Chemical Industry

The Lanchow Institute of Chemical Industry was founded in 1965 and is engaged mainly in petrochemical industry technology research. The bulk of research is conducted in response to the needs of the local chemical works, although the staff occasionally also suggests problems. Both the plant and the institute report to the Provincial

Chemical Industry Bureau in Lanchow. There are 300 scientists and technicians at the institute. It is not a teaching institute and it has no students. All of the staff are college graduates and most of them have been trained as chemists or chemical engineers. The institute would like to have more people trained as analytical chemists, physical chemists, electronic engineers, computer experts. They stated that there simply are not enough trained people in China to fill the needs of industry. Requests for personnel are transmitted to the Education Ministry in Peking, which fills them on a basis of priority relative to other needs.

The institute consists of five divisions: petrochemical industry research division, inorganic and physical chemistry division, analytical chemistry division, environmental protection research division, and anticorrosion research division. The institute is well instrumented and has microreactors, capillary column gas chromatographs, integrators for GC units (Japan), a vacuum evaporator (JEOL) to prepare samples for electron microscopy, an X-ray spectrophotometer (Shimatzu), a reflectance IR spectrometer (JASCO), a mass spectrometer (Atlas-Werke), and a BET instrument for surface ore and pore size measurement. In addition this Institute has the Chinese-made scanning electron microscope shown in Fig. 17.

44. Lanchow Institute of Petroleum Research

The petroleum refinery at Lanchow is associated with two institutes: a Petroleum Research Institute and an Automatic Instrumentation Institute. The Institute of Petroleum Research was built at the beginning of the 1960s and currently includes 330 staff and workers with 150 technical personnel divided into five sections. They are chiefly engaged in research on lubricating oil additives, catalytic cracking catalysts, hydrofinishing catalysts, new oil products, and new refining processes. The main problems arise from production technical problems, but they also work on technical items that fit into their long-range development plans or that are assigned by the Petroleum Ministry. For the past 10 years, they have cooperated with schools, colleges, and other institutes to develop and manufacture metal organic salt detergents and dispersants, oxidation/ corrosion inhibitors, extreme pressure lubricating oil additives, and rust inhibitors. One detergent-type additive being produced is polybutene succinimide with molecular weight of several thousand for use as a viscosity index improver for lubricating oils. They also develop and manufacture bead catalysts for the TCC unit, microspheroidal catalysts for fluid bed catalytic crackers used at other refineries, and catalysts used in lubricating oil hydrotreating. They have developed a Dill Chill Process based on a U.S. journal article (Exxon process). The work is essentially applied, with a very small basic research component.

The project specifically shown involved the examination of succinimides as detergent dispersants. They are also examining primary amines as ashless additives for lubricating oils.

The laboratory is reasonably well equipped with instruments, many of which are of foreign origin such as a Czech apparatus for gas chromatographic coating of capillaries, a British UV spectrometer, an East German vapor phase osmometer, a Japanese vapor phase osmometer, an East German emission spectrograph, a French DTA apparatus, a number of GC units (gas capillary, flame ionization detection), a Zeiss X-ray fluorescence unit to determine metals in catalysts and metals in additives, a Pye Unicam UV spectrophotometer for additive structures, a vapor phase osmometer for molecular weight determination of additives, emission spectroscopy for metal determinations in oil, an elemental analyzer (Carlo Erba), atomic absorption spectrophotometer (SENA), DTA (French) to measure thermal stability of oil additives. Polymer-related equipment includes two vapor phase osmometers (a West German Knauer and a Hitachi-Perkin-Elmer) and a French DTA (Setaram ATDMS) for testing thermal stability of additives.

45. Lanchow Petroleum Refinery

The Lanchow Oil Refinery was the first refinery constructed after Liberation. Before Liberation China only produced a small amount of oil, mainly from the Yümen Oil Field. After 1949 China began to explore for oil and to exploit their oil resources. The Lanchow Refinery was built in the late 1950s with an original design of one million tons/year; it currently has a capacity of two million tons. The oil comes mainly from Sinkiang Province via railroad tank car over a distance of some 2000 km. The refinery was designed and built by the Russians. Since it was an early refinery, it has undergone several expansions and modifications. The essential processing units at the refinery consist of two Thermofore catalytic cracking units; a bimetallic reforming unit; a sulfuric acid alkylation unit; a unit to produce isopropyl benzene for aviation gasoline (H_3PO_4 catalyst); and a heavy oil visbreaking unit. In addition, the refinery has extensive facilities for lube oil processing including propane deasphalting, phenol-extraction, solvent dewaxing using acetone (they plan to switch to MEK) clay treatment, and a Chinese-developed hydrofining process using a catalyst manufactured in Lanchow. The catalytic reforming and alkylation are domestically designed, as are most of the lube facilities. The visbreaking unit is also Chinese-designed via an evolutionary process and is used to convert residual oil into useful products including boiler fuels. The BTX cut from the reformer is not used locally; instead it is distributed nationally. The Sinkiang feed is low in S, N, and metals and has a low paraffin content, so no special precautions need to be taken in shipping in contrast to the highly paraffinic Tach'ing crude.

There are some 700 scientific/technical personnel within the refinery, a reduction from an earlier force of about 1000 people. The rationale for the decrease in manpower is the need for trained personnel at other refineries built in response to the development needs and plans of China. In fact, the Lanchow Refinery, due to its comprehensiveness, is used as a training refinery for personnel and many of its staff are recruited by other petroleum installations. The Lanchow Refinery likes to obtain new personnel who already have some specialized training. The pool of such new people apparently comes from comprehensive training institutions at Tach'ing, Shantung (Shengli Oil Field), and Hupei (Takung Fields). Both Tach'ing and Shantung offer four- to five-year programs in oil field work, refining, chemical machinery, oil exploration, and oil transportation. In Hupei there is a three- to four-year school concentrating on drilling. These programs are under the control of the Petroleum and Education Ministries.

At present the Lanchow Refinery produces gasoline, kerosene, fuel oil, and jet fuel for local consumption. The lubricating oil products are distributed on a wider geographic scale.

46. Liuchiahsia Hydroelectric Plant, Liuchiahsia, Kansu Province

The Liuchiahsia Hydroelectric Generating Station is located on the Hwang Ho (Yellow River) about 60 km southwest of the city of Lanchow. The dam is located close to the town of Hsiaoch'uan and is just north of the Linsha Autonomous District.

The dam is one of a series of four major hydroelectric power and flood control projects on the Yellow River—three of these are located in Kansu Province and one is located further north in the Ning Hsia Autonomous Region. The Liuchiahsia installation includes a dam that is 148 m high and 840 m long, producing a reservoir with a 5.7×10^9 m^3 capacity. The hydroelectric plant includes three generators producing 225 MW each, one generator producing 250 MW, and one generator producing 300 MW. The total station has an output power capacity of 1225 MW, and delivers an average energy of 5.7×10^9

kW hours per year (this output is greater than the total electrical output of China before Liberation). The generating equipment was made in Harbin.

Construction of the dam was started in 1964. The first generator was on line in 1969, and by 1974 all five generators were on line. At the briefing, our Chinese hosts indicated that the dam represented an investment of 630 million Chinese yuan, and that the cost of the power produced was 0.3 cents (Chinese) per kW hour, assuming a lifetime of 100 years for the hydroelectric facility and no interest on the invested capital.

We were told that the Liuchiahsia facility is not the oldest hydroelectric station in the area. The oldest facility is at Yenkuhsia roughly 45 km from this site. This hydroelectric station was put into operation in 1962 and develops 360 MW.

Typical specifications for generators are as follows: The No. 2 generator develops 230 MW at 15,750 V and 50 Hz. Typical rotor speed is 125 rpm with 1662 amp rotor current. Frequency control on the output power was 50 Hz ± 0.5 Hz. The power produced by the hydroelectric plant is distributed as follows:

(a) two 220-kV lines to the city of Lanchow

(b) one 330-kV line to Sian (Shensi Province) through a transmission system that is 530 km long and delivers 400 MW

(c) two 220-kV lines to the city of Hsining. This city is to the east of Lanchow and is the site of the Academy Institute of Saline Lakes.

During the discussion we were told that 80 million cubic meters per year of silt is deposited behind the dam. We were told that the first dam on the Yellow River was completed in 1962 but that a 300-MW coal-fired power plant in Lanchow was built in the 1950s.

APPENDIX E Science Policy and Planning Meeting

A meeting took place on Sunday, May 21, 1978, in the Peking Hotel with a small delegation of Chinese to discuss science organization and planning in China. Representing the Chinese were Ch'ien San-ch'iang, Vice-President of the Academy of Sciences and, at that time, Director of the Institute for Atomic Energy; Liu Ta-kang, Director of the Institute of Chemistry in Peking; Li Su, Head of Chemistry in the Chinese Academy of Sciences; Ch'ien Jen-yuan, Deputy Director of the Institute of Chemistry; Feng Yin-Fu, Deputy Director of the Foreign Affairs Bureau of the Chinese Academy of Sciences. The leader of the Chinese Delegation was Professor Ch'ien San-ch'iang. They had as their official interpreter Li Ming-te. The U.S. Delegation consisted of Seaborg, Baldeschwieler, Bigeleisen, Breslow, Lee, and Suttmeier.

Dr. Ch'ien acted as the spokesman for the Chinese Delegation and took the initiative to give a brief outline of the development of science in China, starting with Liberation. Ch'ien pointed out that in the spring of 1978 a National Science Conference was organized with 6000 participants in all fields of science from all parts of the country. Those attending the conference prepared a National Science Policy. This policy includes a program for the short period from 1978–85. The program has been prepared in broad outline only; the details have yet to be worked out. A broad outline also has been prepared for 1985–2000. At the Conference there was recognition of outstanding work done by scientists in institutes, the army, industry, and agriculture.

Ch'ien indicated that five main sectors had been recognized with responsibility for research tasks: Chinese Academy of Sciences, universities and colleges, industrial enterprises including medical and agricultural organizations, national defense institutes that have research work, and local authorities (this includes the broad masses of peasants engaged in agricultural research). To coordinate the above five units, they have established the Science and Technology Commission.

The Academy of Sciences has been given lead responsibility for research in the natural sciences. The Academy is organized into five units: mathematics, physics, and astronomy; chemistry; biological sciences; earth sciences; and new technology—computer sciences, semiconductors, etc.

As a result of the proposals made by the various delegates to the National Science Conference, priorities for areas of science and technology were developed. The highest priority areas include: (1) agriculture, (2) material sciences, (3) energy resources (including nuclear, solar, fossil fuel, etc.), (4) computer science, (5) space science, (6) laser science, (7) high-energy physics, and (8) genetic engineering. The rationale behind these eight programs is as follows: The first three—agriculture, material sciences, and energy resources—are designed to meet applied programs and current needs. The next three areas—computer science, space science, and laser science—are designed to bring Chinese science and technology up to that of the advanced countries. The areas of high-energy physics and genetic engineering were selected as two prestige areas in which Chinese science should be up to that of advanced world levels by the year 2000.

There was a significant debate as to why the general area of health and medicine was not on the list of eight. Ch'ien said that this area had been considered, but it did not make the final list. We then discussed in detail why high-energy physics received such high priority. In particular, Seaborg pointed out that this was an expensive area and that concentrating a large amount of resources into this research might deny the country many research opportunities in other areas.

The next topic discussed was how a working scientist gets a new idea or research program into the system. Ch'ien said that the scientist would discuss the project with the director of his institute. It would be reviewed there, and if acted on favorably by the institute, then funding would be requested from the Chinese Academy. If the proposal did not get favorable response within the institute, then the individual scientist could appeal through the system all the way to the top. Ch'ien pointed out that getting all the way to the top very rarely happened, but that it is possible.

The discussion concluded with a brief outline by Baldeschwieler of the general scheme of science organization in the United States. This was followed by a presentation by Breslow, who talked about the role of the individual investigator, peer review, and individual funding. Finally, Bigeleisen gave a brief summary of the role of the National Academy of Sciences and the National Research Council. A number of recent studies within the Research Council were mentioned, including the halocarbon study, the pending saccharin study, the study on energy and climate, the planning for the large space telescope, and the synchrotron radiation facility.

INDEX

U

V

W

X

Y

Z